Siscia, Pannonia Superior

Old and new finds

Remza Koščević

BAR International Series 2461
2013

Published in 2016 by
BAR Publishing, Oxford

BAR International Series 2461

Siscia, Pannonia Superior

ISBN 978 1 4073 1069 5

BAR Publishing is the trading name of British Archaeological Reports (Oxford) Ltd.
British Archaeological Reports was first incorporated in 1974 to publish the BAR
Series, International and British. In 1992 Hadrian Books Ltd became part of the BAR
group. This volume was originally published by Archaeopress in conjunction with
British Archaeological Reports (Oxford) Ltd / Hadrian Books Ltd, the Series principal
publisher, in 2013. This present volume is published by BAR Publishing, 2016.

Printed in England

BAR
PUBLISHING

BAR titles are available from:

BAR Publishing
122 Banbury Rd, Oxford, OX2 7BP, UK
EMAIL   info@barpublishing.com
PHONE   +44 (0)1865 310431
FAX   +44 (0)1865 316916
www.barpublishing.com

TABLE OF CONTENTS

LIST OF MAPS

LIST OF FIGURES

OLD AND NEW FINDS

Introduction

More than a decade and a half ago, a volume was published in this series with an exhaustive overview of the historical facts and archaeological finds of Roman Siscia (Siscia 1995).

In the meantime, the quantity of finds has increased considerably and their significance now requires a partial publication of earlier finds and, naturally, the full publication of the new finds from systematic excavations.

The first section of this volume presents certain older, elsewhere published or previously unpublished finds from museums and collections, included on the basis of their importance in relation to Siscia.

The second part of the volume encompasses recent finds from the excavations performed in 2003 at the site of St. Quirinus in Sisak.

The finds are analyzed in reference to material published fifteen years ago, to enable consideration of their interrelations, numerically supplement the same kinds of forms, and draw attention to certain less common or rare items, with the aim of expanding the cognitive perspective about this city and its products, which experienced an uninterrupted span of activity from prehistory to the present day.

Data known to the present about Siscia are repeated here in an amount reduced to the framework of the general, as necessary for following and understanding the text.

Map 1. The position of the city of Sisak on the map of Croatia (Lolić 2007: Fig. 1)

1

Map 2. The position of Siscia within the borders of the Roman Empire (Burkowsky 1999: Fig. 9)

Present-day Sisak (Map 1.) has existed with a continuity of over two millennia in the same place, which has even preserved the original name. The modern town is almost entirely superimposed on the Roman city, meaning that large-scale excavations cannot be performed and investigation is limited to urban fragments in rescue interventions depending on the city's municipal projects.

Local interest in the antiquities of Siscia already existed at the turn of the 18th century, and the circumstance that the modern city still lies on the same part of the waterways has resulted in an abundance of ancient material removed from the Kupa River (*Colapis*), particularly during the dredging of its bed in various campaigns at the beginning of the 20th century.

Siscia

Siscia, which in terms of its favorable geographic position (Map 2.) was among those Roman cities that were relevant for the strategic aims of the Empire, was noted on the Tabula Peutingeriana (Map 3.), and its historical fate can be traced through information in ancient written sources. In regional frameworks, it was one of the most important places of the Roman province of *Illyricum*, which incorporated the later independent provinces of Pannonia and Dalmatia.

Siscia was preceded by the pre-Roman settlement of

Segestica, or *Segesta*, which probably originated in the 4th century BC, at the time of the Celtic invasion of Greece, which resulted in the conquest of the Sava River basin and part of the Danube basin. Information is scarce about this settlement: its name can sporadically still be found in the last third of the last century BC, but after that Roman authors no longer mention it. The ethnicity of its inhabitants (*Segestanoi*) remains one of the mysteries of this generally very poorly known place, but it must have been some Celtic-Illyrian assimilation, arising from intermingling with the local autochthonic tribal communities of the Kupa River basin (Siscia 1995, Map 4).

Segestica was located on the right bank of the Kupa River – opposite *Siscia*, at the site of Pogorelec (Fig. II), where remains of a Celtic settlement were registered. However, until more thorough investigation is undertaken, concrete knowledge about the character of the site is unavailable and its relations with *Siscia* remain unclear.

Siscia was granted the status of a Roman colonia in the reign of Vespasian (*colonia Flavia Siscia*), which was again confirmed in the reign of (*colonia Septimia Siscia Augusta*).

After Trajan's division of Pannonia into two provinces, *Siscia* became a part of *Pannonia Superior*, and after Diocletian's separation of the southern part of Pannonia Superior into a separate province – *Provincia Savia* (Map

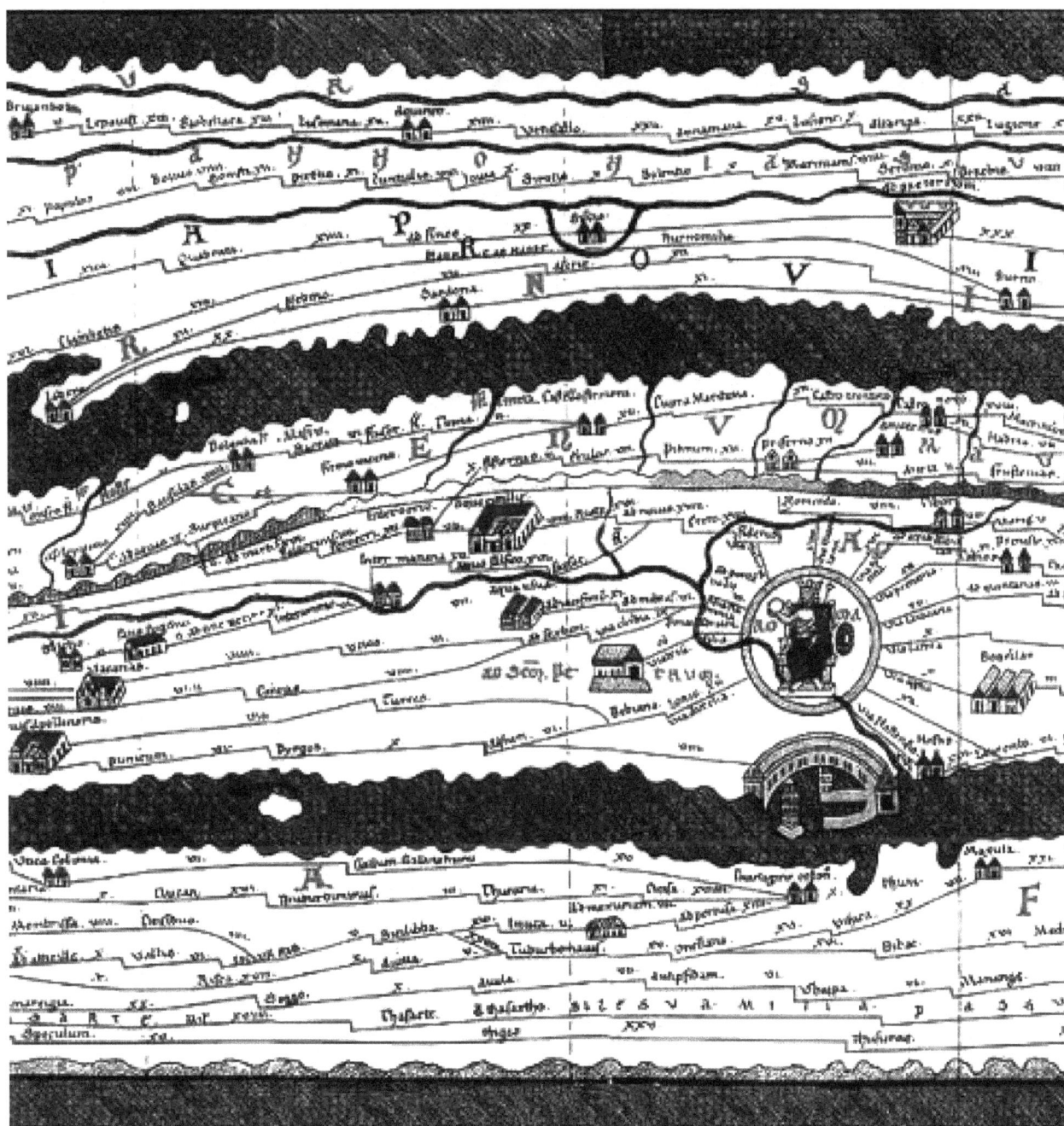

Map 3. A segment of the Tabula Peutingeriana (www.madzag.hu)

4.), *Siscia* became its main city, with all the prerogatives and facilities of a capital. It contained, among other things, the provincial treasury and an imperial mint, a customs station (*dispensator provinciae Pannoniae*), the mining administration (*praepositus ferrariarum*), and a cult centre (*sacerdos provinciae Pannoniae Superior*). *Siscia* was located on one of the largest road junctions (*caput viarum*), and also had a river port. It counts among cities that evolved from military camps. A permanent military garrison was stationed in *Siscia* as early as the end of the last century BC (Fig. I.). It is thought that the *Legio IX Hispana* and the reserve troops (*vexillarii*) of *Legio XIV Gemina* were stationed in the *Siscia* camp.

In the last few decades, the urban topography has been supplemented in several places by discoveries of fortifications and residential architecture: the *horreum* and part of the city walls with a tower, a *villa urbana* with heating, plumbing, and sewage system, a wooden structure with piles (Fig. II.), and the foundation remains of other structures (Pregled 2000: 34-39, 42-46). The exact position of the military camp has not been discovered to the present.

Siscia had four cemeteries (Fig. III.), and perhaps a fifth, but only for the southern one was it established that it lay along the exit road, near the walls but on the outside. All four cemeteries have been partially or completely excavated,

3

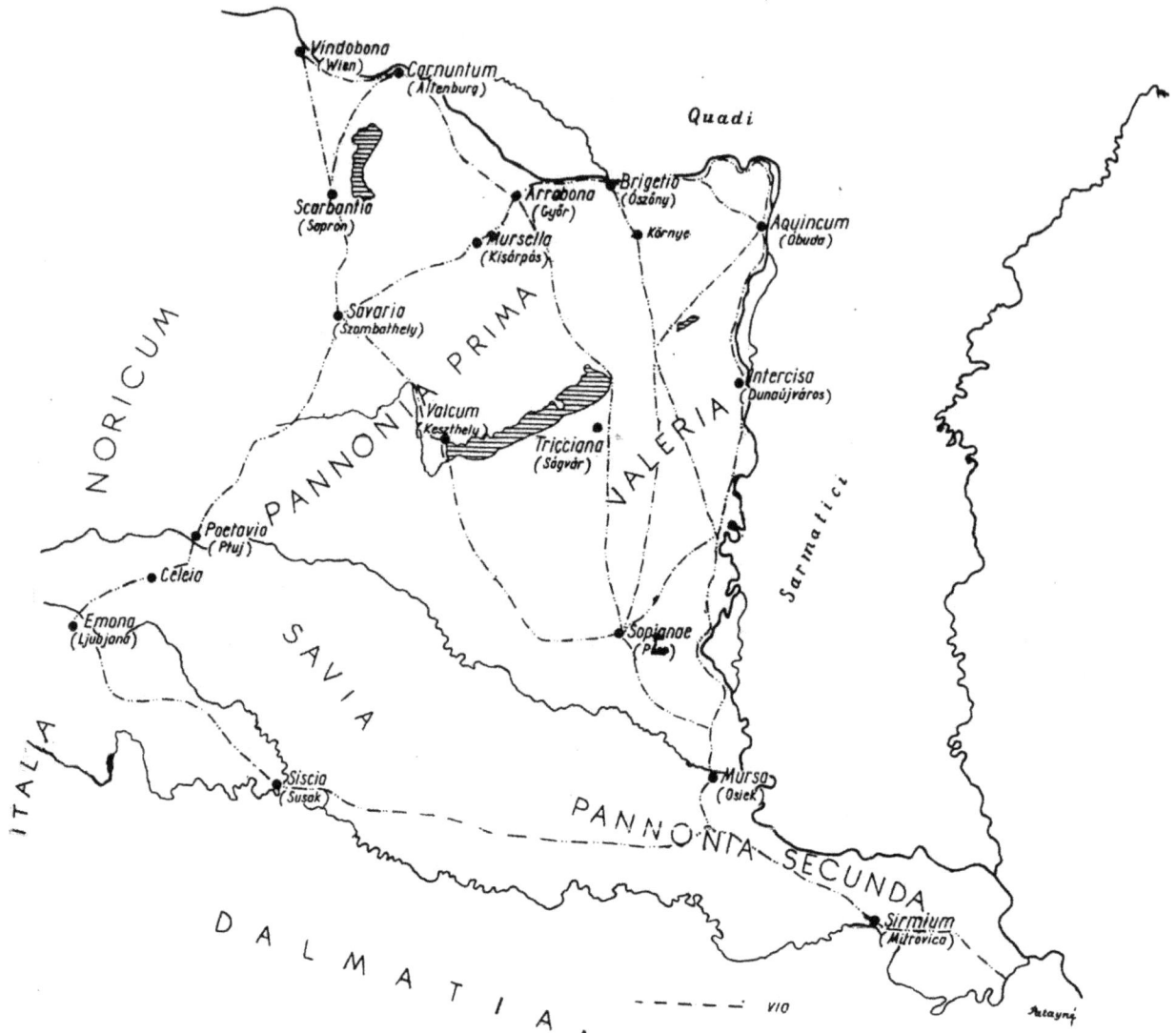

Map 4. The division of Pannonia into four provinces (Burger 1966, Fig. 1.)

while the contents of the grave inventories have only been published for the southeastern necropolis (Wiewegh 2002: 6-12).

In a span of almost two centuries of collecting mobile archaeological material, Siscia has produced an abundance of small finds that, in the absence of more monumental and better preserved static remains, have considerably supplemented the epigraphic material. The particularly large number of metal finds and the fact that some of them were found in states showing various phases of the production process illuminate knowledge of this city as a powerful production centre of the Roman world.

The foundations of the more massive *Siscian* production have their roots in the military camp. Military workshops supplied the garrison with equipment, primarily weapons, but also parade attire elements, as well as certain special categories of small necessities related to the military administration.

Siscia was supplied with raw metal from the nearby mines in the valleys of the Japra and Sana Rivers, where the ore was surface mined (Fig. IV.). The excavated ore was smelted on the spot and poured through tubular ceramic channels, and the cylindrical, rough, and porous raw castings (Fig. V.), were subsequently melted in the smelting works into a kind of ingot suitable for transport. The permanent demand not merely for weaponry, but also various craft and agricultural tools, meant that the need for iron was evidently considerable. However, a considerable part of the production in Siscia was in bronze, and also in lead, but the expected discovery of traces of alloying bronze have not been confirmed, and remains of copper ore exploitation at mining sites throughout the broader area that gravitated towards Siscia have not been documented.

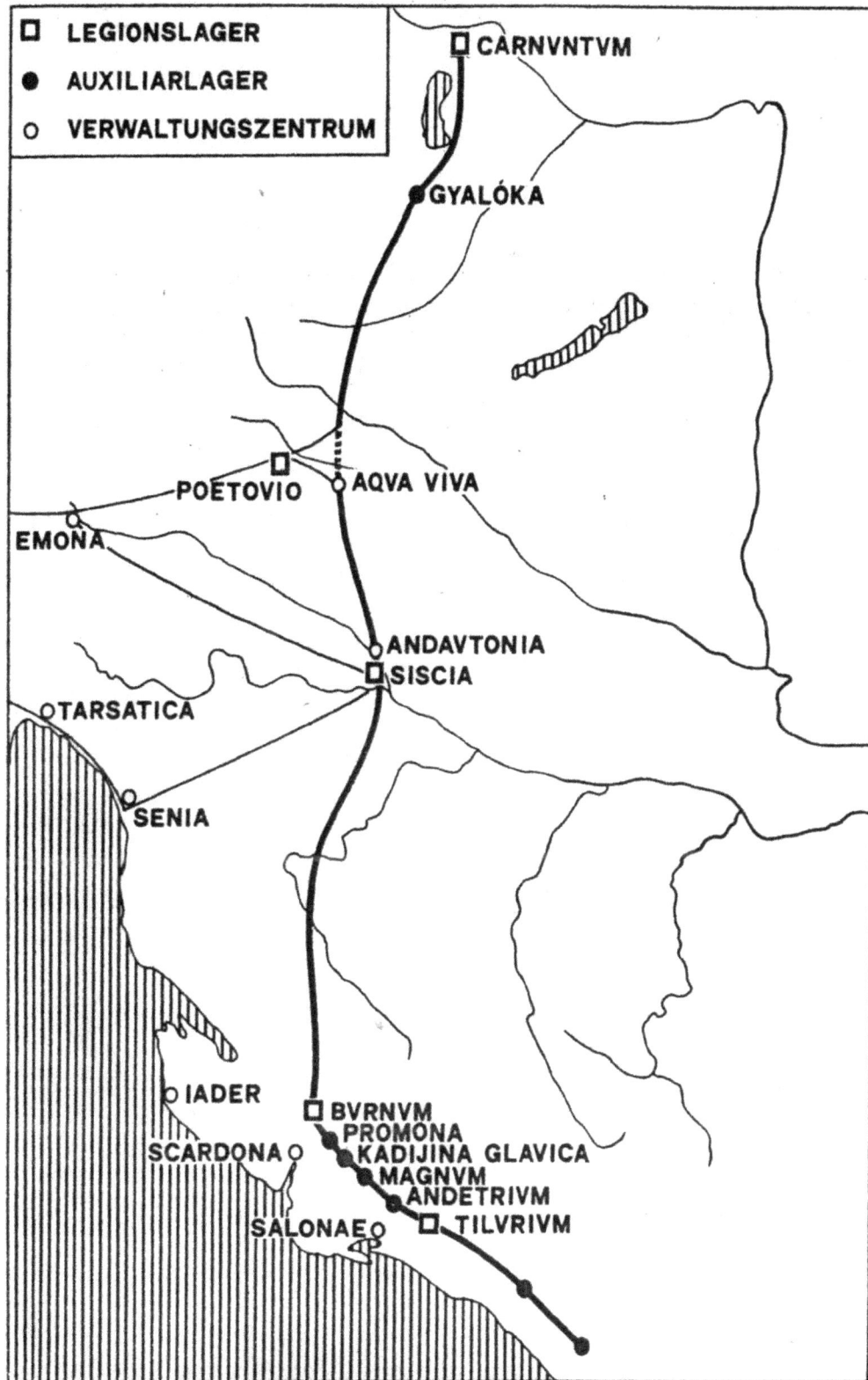

Fig. I The positions of legionary fortresses in Dalmatia and Pannonia at the beginning of the new era (Šašel 1992, Fig. 1).

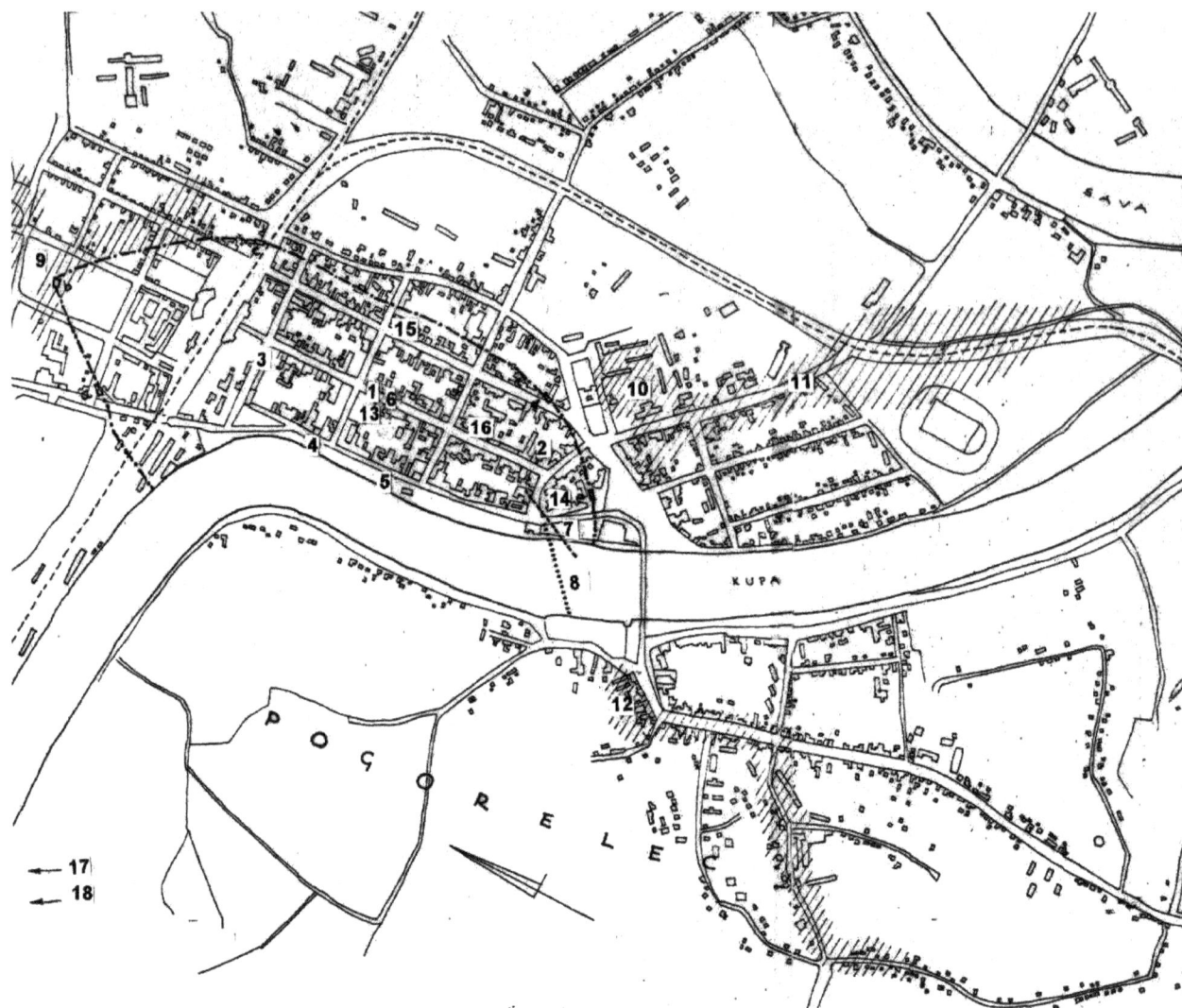

Fig. II The Roman nucleus of Siscia on the plan of the modern city (Faber 1973 – Vrbanović 1981):
1. Finds of amphorae in a "merchant's house"; 2. Late Roman building; 3. Stone building; 4. Baths; 5. Forum or Capitolium; 6. Basilica; 7. Route of the Roman sewer; 8. Water pipeline and Roman bridge; 9. The Roman cemetery next to the Church of St. Quirinus; 10.-12. Roman cemeteries; 13. Wooden structure with piles, a well, and a glass-working furnace;14. Horreum and part of the Roman city walls with a tower; 15. Villa urbana with a portico, heating, water supply and sewage system; 16. Hypothesized trading-craft facility; 17. Traces of the Roman bridge at the Odra River; 18. The influx of the Odra into the Kupa.

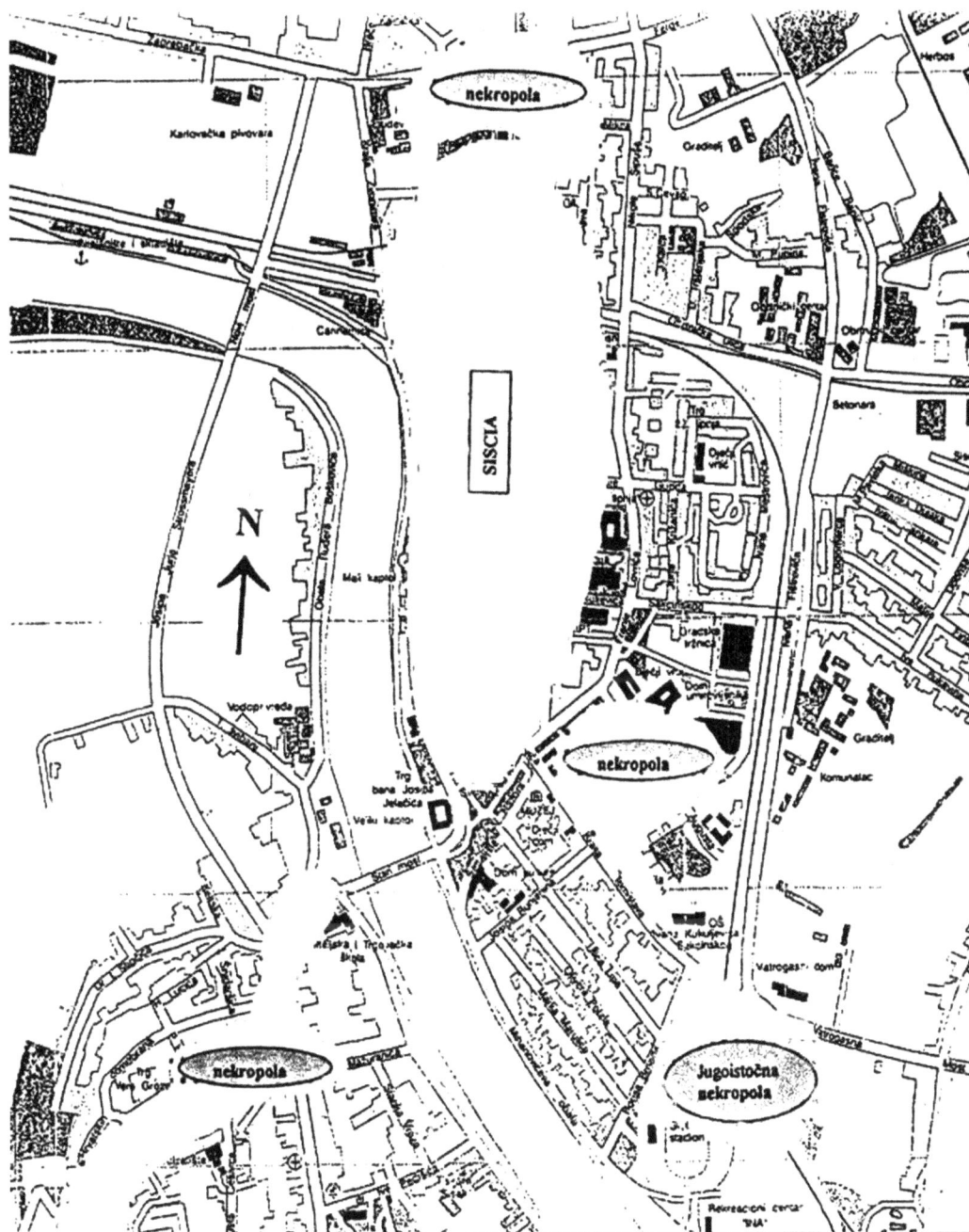

Fig. III The cemeteries of Siscia (Wiewegh 2002: Fig. 3).

Fig. IV Surface extraction of iron ore at ore deposits in the Japra River valley: the sites of Maslovare and Čelopek (photo: R. M.).

Fig. V Clay tubes for discharge of molten iron (photo: R. M.).

On the basis of archaeological facts that have been recognized and evaluated by professionals, the manufacture and storage of the metal goods in Siscia took place somewhere on the right bank of the Kupa, in the sector near the Roman bridge that linked the two opposite banks (Fig. II.), where the mint for coins must also have been located.

Part I

Older Finds – Published and Unpublished

The finds are kept in the Municipal Museum of Sisak (further: GMS), the Archaeological Museum in Zagreb (AMZ), in the Zagreb collection of Matej Pavletić (Pavletić Coll.), the Sisak collection of Marko Golan (Golan Coll.), and in one anonymous numismatic collection in Zagreb (Numis. Coll.).

Published finds

In contrast to the Golan Coll., which can definitely be attributed to Siscia, the extensive and very rich Pavletić Coll. which among other things includes a very important earlier known male head fashioned in the Celtic manner (Tragovi 2003: 82, cat. no. 147) attributed to the Siscian ceramic sculptural workshop (Siscia 1995, Pl. 11, 59-65, Pl. 12, 66), a miniature sculpture of Jupiter (Tragovi 2003: 85, cat. no. 153) of the type published in Siscia 1995: Pl. 12, 68, but with a somewhat different arrangement, and the figure of a female deity (Tragovi 2003: 89, 90, cat. no. 164, 165) from the series of lead cult figurines of Siscian production of the type published in Siscia 1995: Pl. 51 etc.), contains relatively few objects with confirmed proveniences of Siscia as their site of discovery. This particularly refers to the lead lunular pendants and certain forms of fibulae, for which the site of discovery is not noted (Tragovi 2003: 95, 102, 103, cat. no. 192, a-c, cat. nos. 223-226), although their very appearance, typical for the Siscian region, would speak in favour not merely of their discovery there but also their provenience in Siscian workshops.

The most interesting item from the Pavletić Coll. is a miniature bronze figurine of Artemis-Diana (Tragovi 2003: 87, cat. no. 157), dated to the 2nd-3rd centuries. She was depicted according to the classic Greco-Roman cult standards (Fig. VI.) in terms of the clothing and the hind and dog accompanying her. If the attributes had been preserved in her outstretched hands had been preserved and if they were a pair of torches, then it could have been a syncretic representation of two classical female deities, the other one being Hecate. However, since in the current state of preservation the figurine has empty hands, the main attribute then becomes the crown on the head, which has a unique appearance. It contains evident elements not merely of a lunar cult, but also of a fertility cult of universal character, and under its auspices was presented a probably original personified version of some pre-Roman belief, perhaps even with Celtic religious roots, which was externally adjusted to a Roman interpretation.

Fig. VI A bronze figurine of Artemis with a distinctive crown on her head (Tragovi 2003: 20, Fig. 6).

And both originally empty-handed bronze figurines of Mercury from the Golan Coll. (Fig. VII a, b), exhibit a deviation in the design concept and its execution (Koščević 1999, no. 1, 2). The depictions of this otherwise most popular and most frequently artistically represented Roman god appear in a variety of iconographic images. For the Egyptian Thot, the creator of writing and hermeneutics, the characteristic attribute was an ibis feather. His Hellenistic counterpart – the Greek Hermes, first as the messenger of

Fig. VII A bronze figurines of Roman Mercury without his primary attribute (Koščević 1999, Fig. 1, Fig. 2).

the gods and psychopompos, and only later as the protector of merchants, most often wears boots with wings or snakes (kerykeion). However, the Roman Mercury is only very occasionally represented with other attributes, such as reptiles: turtles, lizards, chameleons, which in their entire symbolism negate the status of a practical god, discreetly indicating the original properties of his underground version – chtonios. In the Roman edition, he is almost exclusively the god of trade (as represented by the root of his name) and financial gain, depicted with the mandatory attribute of a purse (marsupium), which was added only in the interpretatio romana.

The first figurine (Fig. VIIa), represents the standard, previously known (among Siscian miniature sculpture) Roman depiction with a winged cap and a caduceus in the left hand and a cloak wrapped around the same arm (Siscia 1995, Pl. 13, 74); the figurine is shown in the classic posture of the far distant Praxiteles model, but in a counter stance, with the weight on the left leg and with an open right hand missing the purse.

The second, double figurine (Fig. VIIb) offers a generally very rare depiction of Mercury with a companion. The figure of Mercury is somewhat leaning on the right leg, and has only secondary attributes: a petasos with wings and a cloak

over the right shoulder. The identification of the female figure in the long flowing clothing is difficult because of the lack of any attribute or symbol. Her identification with Fortuna seems most plausible considering the identical nature of both deities as the bearers of prosperity, and also on the basis of the fact that in the home lararium, Mercury was often associated with Fortuna.

However, since both depictions – both the single Mercury and that paired with a female deity – contain very few Roman elements, and the proportions of the figures and particularly the presence of a naïve nuance typical for the Celtic treatment of physiognomy would prevail in favour of the latter, the equal possibility remains that the female figure could represent the Gallic Rosmerta or Maia, or even some local, as yet nameless deity. Both figurines should be defined as depictions executed in a Roman-Gallic interpretation, with a more strongly expressed Celtic component on the second example. They represent the modest products of some probably Siscian production centre from the period of the 2nd-3rd centuries.

The bronze ithyphallic figurine of Silvanus from the Golan Coll. (Fig. VIII) exhibits a quite original version (Koščević 2002, Pl. 1, 1), which is based on the known iconographic model of a combination of the grotesque dwarf Egyptian

Fig. IX A bronze bust-weight (Koščević 2002, Pl. 1, 2).

Fig. VIII A bronze figurine of an unusual Silvanus-Bess (Koščević 2002, Pl. 1, 1).

Bess in a Roman interpretation and the Italic Silvanus. Their syncretism is based on their common characteristics: the first as an indispensable participant in wedding ceremonies and those devoted to Isis, and the second as a participant in the Dionysian processions. Unlike previously known examples in this form, which were most often vessels, this specimen represents a cult figurine. However, it is lacking essential properties of its iconographic model, or perhaps it was deliberately imbued with a different conception of both these deities belonging to a lower rank. The superficial, naïve, and almost caricature-like concept of exaggerated erotic charge and its partial modelling and extremely robust execution define it as a below average product from the middle Imperial period, which could have been made in any workshop, even the least well equipped, perhaps even in Siscia itself, where evidence already exists for the practice of the cult of Isis.

A rare example of depiction is shown on a large bronze weight for a steelyard or scales from the Golan Coll. (Fig. IX), which does not fit among the usual repertory of the known figural weights with depiction of civil, military, imperial images and figures with a mythological or religious character. The bearded bust of non-Roman appearance (Koščević 2002, Pl. 1, 2), both in terms of the facial features and the hairstyle as well as the clothing, depicts a barbarian

– a Dacian or Scythian. A confusing element is the band tied around the head, which can be interpreted in two ways: it can either be considered a banal detail of eastern apparel or as a divine symbol – a taenia, which would suggest that the depiction was probably of Jupiter in a Dacian interpretation.

Significantly contributing to the quantum of Siscian fibulae is a group of them from the Golan Coll., containing examples of bow fibulae of the Aucissa, Kraftigprofilierte/ Profiled, Knie/Knee, Trompetten/Trumpet, and Zwiebelkopf/Onion head types (Koščević 2000/1, Pl, II, 26 – 28, Pl. III, Pl. IV), which are congruent to such previously known types (Siscia 1995, Pl. 35, 348, 352, Pl. 36, 356, 360). The only as yet unrecognized form within the corpus of Siscian fibulae is represented by a 13.5 cm long fibula of mixed character (Fig. X), which in addition to elements of the Germanic Boierspangen (Siscia 1995, Pl. 35, 347) also contains an archaic line of the bow and form of the foot that can be found on late La Tène forms and chronologically correspond to the early 1st century (Koščević 1980, Pl. I, 2, 4).

The publication of a group of identical examples of annular fibulae (Fig. XI) in the collections of the GMS (Wiewegh 2003), increased the quantities of known annular clasps with open and closed rings of the type published in Siscia 1995, Pl. 26, 218, Pl. 40, 406.

According to available publications, in comparison with the Siscian finds made from other materials, glass objects appear to be relatively poorly represented, despite the already proven evidence about the production of certain glass items in Siscia (Koščević 1996: 84). Along with the predominant balsamaria of tubular and similar forms, the GSM also contains beakers, circular and square flasks with a handle, bowls, gutus, but also fragments of millefiori vessels as well as aryballoi (Burkowsky 1999: 67; Wiewegh 2002: 10-12, 15, cat. nos. 167-206), including an example with a completely preserved bronze handle (Fig. XII).

Within the very small group of glass from the Golan Coll., composed of the inevitable tubular balsamaria, conical beakers with a ringed foot, olla cineraria, plates in the form of patera (Koščević 2003: Fig. 1, 2), some rarer forms of

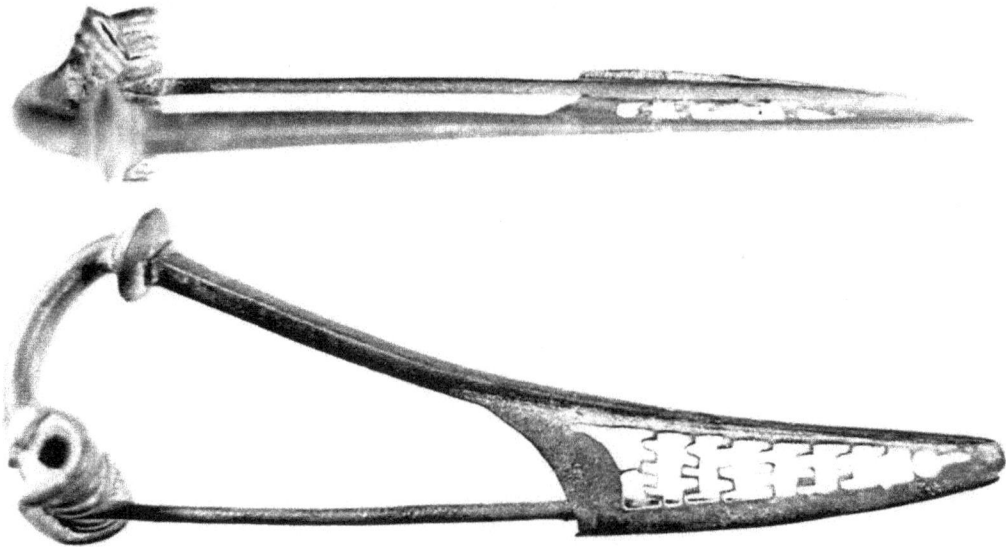

Fig. X A large bronze fibula of mixed type (Koščević 2000/1, Pl. II, 25).

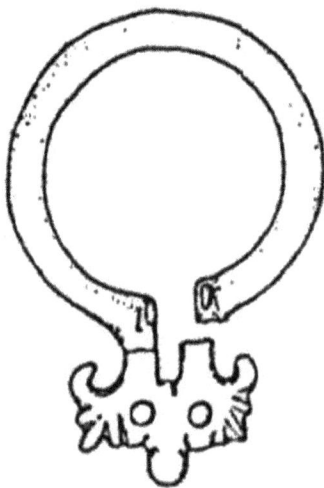

Fig. XI A bronze annular clasp with a pierced extension (Wiewegh 2003, Pl. I, 1-4).

Fig. XII An ariballos with a bronze handle (Wiewegh 2002: Fig. 12).

balsamaria were also found, probably from around the 1st century (Fig. XIII), which have no direct analogies elsewhere or only approximate ones.

The bronze pin with a polyhedral platelet from the Golan Coll. (Fig. XIV), with a male or female head facing left stamped in relief (Koščević 2001/1, Pl. I, 10), so far is the only example with figural decoration within the previously known group of 16 examples with oval, lunular, pelta, and swastika shaped platelets, as well as those in the shape of a key (Siscia 1995, Pl. 41, 417 – 431), dated to the late 2nd and into the 3rd century, which appear extremely rarely outside of Siscia.

Also almost unknown outside of Siscia are the lead pendants with figural depictions stamped together with an edge decoration, which include: a bust of Mercury facing right, a figure of Victoria facing right with a wreath, and a figure of Mercury facing left with a purse but also with a wreath, known to date only from these three examples (Siscia 1995, Pl. 53, 568 – 570). The fourth lead example

Fig. XIII Examples of balsamaria with a funnel neck and a globular body (Koščević 2003, Fig. 2, a,d).

from the Golan Coll. (Fig. XV) is completely identical to the last listed example, including the dimensions (Koščević 2001/1, Pl. III, 22) and it can be determined as having been stamped from the same die and attributed to Siscian lead production of the 1st – 2nd centuries.

There is a reasonable probability that these pendants were actually the lids of containers to protect seals: indirect evidence for this possibility are the previously known Siscian seal-box containers made of lead, including examples with the identical leaf-shaped form as these four examples (Koščević 1991: 31, cat. no. 71, 72, Pl. II, 37).

The case is somewhat the opposite for the bronze example of a seal-box from the Golan Coll. (Fig. XVI), which is the only one of four examples of such containers used to protect seals stamped in wax to have been completely preserved, i.e. both box and lid. This is an example of the leaf-shaped type with an enamelled lid, also known from Siscia in a variant with an internal partition in the form of a reversed pelta (Siscia 1995, Pl. 34, 337, 338). A second variant with a heart-shaped internal partition on the lid, to which this example belongs, refers to the most widespread form among seal-boxes decorated with enamel, known to date from various sites in a quantity of 16 identical examples.

This shape had not yet been noted among the abundant Siscian groups of such boxes of various typological styles (Koščević 2000/3: 14).

In addition to the great amounts of clay lamps of almost all known types (Siscia 1995, Pl. 10) and a more modest quantity of metal examples (Siscia 1995, Pl. 20, 165-170), other types of Roman lighting from Siscia are much more poorly represented. A high quality cast example of a simple three-legged candelabra with elegant lines from the Golan Coll. (Fig. XVII), was originally covered with a white metallic coating as an imitation of silver plating, which was a quite common process of decoration in the early and middle Imperial periods. It could be used as a holder or stand for oil lamps, or as a true table or altar candlestick with candelae (Koščević 2002, Pl. 4).

An example of a two-part bronze pendant in the AMZ (Fig. XVIII), which was previously only superficially published as an illustrative design (Strena Buliciana 1924: 236), represents a significant find from Siscia, primarily because of the lunula, which are considerably lacking on the known forms of pendants and fittings from military equipment, but also because of the now empty circular hollows originally filled with figural elements in a white metal, such as can be

Fig. XV A lead pendant with a depiction of Mercury (Koščević 2001, Pl. III, 22).

Fig. XIV A bronze pin with figural relief decoration (Koščević 2001, Pl. I, 10).

Fig. XVI A bronze seal-box with a heart-shaped partition for an enamel fill on the lid (Koščević 2001, Pl. II, 12).

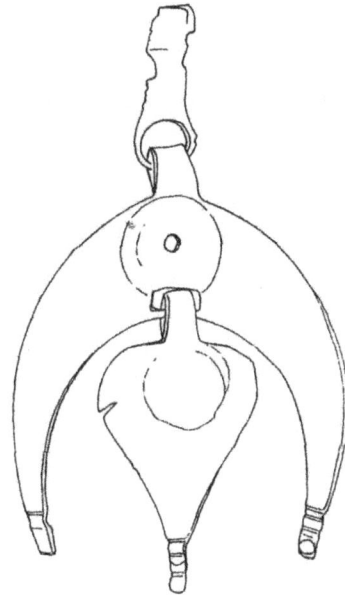

Fig. XVIII A bronze lunular pendant with a leaf-shaped hanging element with a lost original decoration (Strena Buliciana 1924: 236).

Fig. XVII A three-legged bronze candelabra with panther protomes (Koščević 2002, Pl. 4, Pl. 5.).

seen on other forms of pendants from military equipment (Siscia 1995, Pl. 27, 234, Pl. 19, 136-146).

As is shown by the leading types of bronze pendants from early Imperial equestrian equipment produced in Siscia (Fig. XIX, a-d), forms derived from leaf shapes were far more widely represented (Siscia 1995, Pl. 27, Pl. 28), although their origins, as well as their ornamentation, were based on a more complex floral and animalistic symbolism. If we look at material from other sites that evolved from legionary camps along the Danube, Drava, or the so-called Illyrian limes (Fig. I), sporadically we will come across originally identical forms, but with different conceptual characteristics and, naturally, different workshop origins (Tekija 2004, Cat. Nos. 2, 4, 8, 9; Poetovio I 1999: 67, Fig. 54; Burnum 2007: 28, fig. 21). On the other hand, however, the Siscian finds display the greatest similarity, not merely in forms but also appearance, to material discovered at Baden and Windisch (Unz 1972, Abb. 5, 46, 47, Abb. 6; 1974, Abb. 12, 149), where all four typical forms for Siscian production were represented in large quantities (Fig. XIX).

The only published lead example of a so-called Bleibulla with a figural depiction (Fig. XX), has thrown some little light on a group of several dozen unpublished examples of a very interesting category of lead seals with a small tubular channel for a cord on the back and with a relief depiction of busts and profiles – including imperial ones – stamped with a die, with the appearance of a miniature (Gorenc 1979-1980).

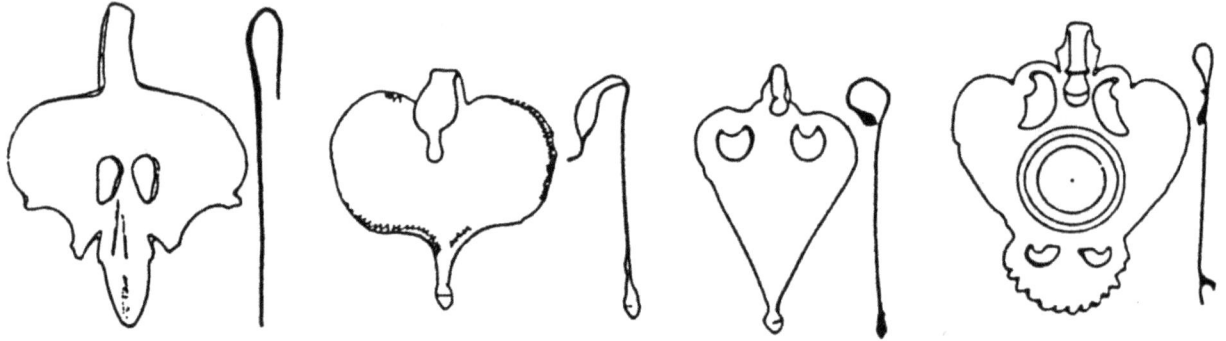

Fig. XIX a-d. The main forms of pendants of Early Imperial equestrian equipment with Siscian workshop provenience (Koščević 1991, Pl. XIV, 208, 213, 215, Pl. XV, 220).

Fig. XX A lead bulla with a figural depiction in relief (Gorenc 1979-1980).

Unpublished finds

Among the various categories of lights, a rare kind of find is represented by the metal sections of Roman lanternae (glass lanterns). Two such bronze fragments come from Siscia: a deformed and damaged cylindrical burner with a widened base, an inverted and partly broken off bottom, with engraved double horizontal lines on the exterior side, and a foot with a spherical base with a notched ring-shaped furrow on the upper side (Fig. XXI, a: AMZ, inv. no. R-5807, h. 34 mm, dia. 50 and 40 mm; Fig. XXI b: GMS, inv. no. 687, h. 17 mm, dia. 13 mm).

Attention should also be paid here to three Siscian bronze hooks, found together (AMZ, listed under the same inv. no. R-6284/1-3, h. 56, 62, 82 mm), two of them cast and modestly decorated with segments of horizontal folds and with a circular opening on one end and a biconical button on the other end (Fig. XXII), which certainly served for hanging lanterns. Although it is hypothetical, this group find could signal that the hooks had belonged to a stand (lychnucus) where several lamps were suspended at the same time to intensify the lighting.

Fig. XXI a, b. A bronze burner and bronze foot of a lantern (photo: Z. G.).

Fig. XXII Bronze cast hooks (photo N. M.).

Fig. XXIII Example of Siscian lead labels (photo: N. M.).

The earlier well-known lead labels, for which Siscia is in general the major site of discovery with over 1000 examples (Siscia 1995: Pl. 54, 587, 588), have been joined here with several unread examples from the Numis. Coll. (Fig. XXIII), but only in terms of registering them with a photograph, and not their analysis. These simple plates with graffiti inscriptions from other sites most often bear incised data about foodstuffs (vinum piperatum, garum, asparagus) or attire (sagum, paenula) products.

Out of the numerically impressive group from Siscia removed from the Kupa River, two examples were published at the beginning of the 20th century that were defined as notes about monetary amounts (Brunšmid 1901:

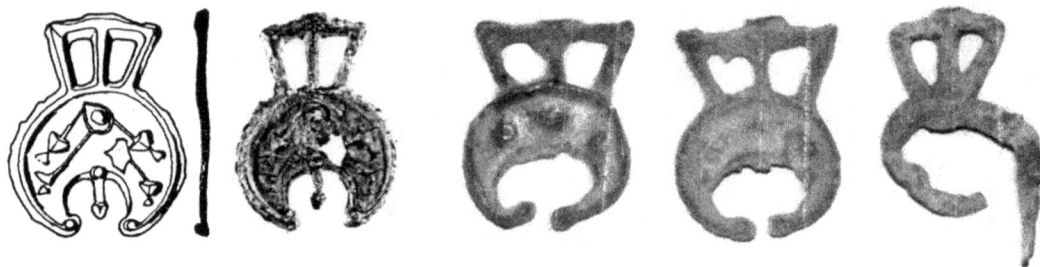

Fig. XXIV a - d. Lead lunular pendants manufactured in Siscia (photo: N. M. and Z. W.).

124, 125, fig. 87). In the middle of the same century, another 21 examples from Siscia were published that were kept in the Hungarian National Museum, defined as being marked for the storage of raw wool that had arrived in Siscia from Istria and Dalmatia (Mócsy 1956: 104).

The quantities in which these lead bullae and labels (dated from the 1st to 3rd centuries) were found would indicate their possible connection to the existence of a customs office (portorium publicum) in Siscia, in which both these categories of administrative supplies were produced.

The group of previously known and often discussed (Koščević 1990; 2001/2) lead lunular pendants (Siscia 1995, Pl. 53, 574-584), dated to the second half of the 2nd century and the 3rd century, are now joined by another four newly documented lead examples with a crescent-shaped platelet (Fig. XXIV, a: Numis. Coll., dia. platelet 23 mm; Fig. XXIV b - d: GMS, no inv. no., dia. platelet ca. 20 mm), increasing the number of examples of these pendants with a Siscian workshop provenience to 40. The first pendant represents the most intriguing variant with a complex decoration, and the other three smaller examples – the last one unsuccessful and discarded – belong to the same but simplified variant decorated with three granules, which is in general the most common type. The earlier treatment of these aniconic pendants as requisites within the cult radius of some female deity has not changed (Koščević 2001/2: 148), as information about this still remains firmly enclosed in the pendants themselves, although it might be exposed to some extent by the crown on the head of the small goddess from the Pavletić Coll. (Fig. VI).

The last example listed here, a silver medallion with a relief figural depiction (Fig. XXV: Numis. Coll., l. 21 mm), is interesting from the standpoint of its possible double function. It was fashioned from thin silver sheet metal, including a practically designed loop for suspension, made by simply bending the cut out band onto the front side. It is the fifth example within a small series of the so far known four bronze specimens with the impressed busts of Jupiter/Serapis, Minerva facing right, and the figure of Fortuna with a cornucopia and a rudder, as well as a charioteer (auriga) with a quadriga (Siscia Pl. 46, 487-490). The depiction on the silver platelet is: a seated female figure in left half-

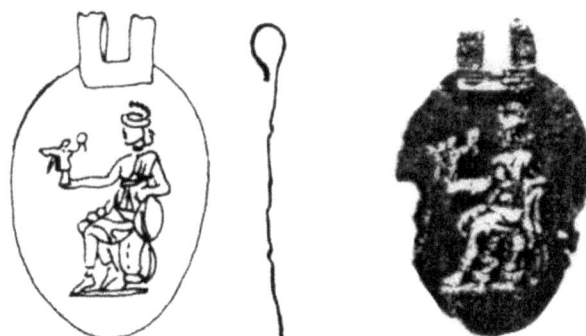

Fig. XXV A small silver medallion with a miniature relief depiction of Dea Roma (photo: N. M.).

profile, in a short chiton with the right side of the chest bared, with a crested helmet on the head, and a winged Victoria with a wreath in the outstretched right hand; the top of the right foot touches the support, and the left leg, clothed above the ankle, has the left foot on the ground as marked by a horizontal line; to the left of the figure at thigh height a circular shield is leaning, with armour underneath it. The figure represents a depiction of Dea Roma in a version of the type that appeared as early as the Republican period, became common at the beginning of the first century and continued throughout the entire Imperial period, while its iconographic origins date back to the Hellenistic image of Athena Nikephoros as a prototype. The depiction represents a numismatic and gemological motif, and can often be seen on Imperial coinage (Titus, Nero, M. Aurelius) and on intaglio items from the 1st century (Sena Chiesa 1966: 250, Pl. XXXIII, 646, 647).

In the first publication of these medallions, they were classified in terms of form among pendants (Koščević 1991: 42, Pl. XII, 173-176), and such a definition can continue to be tolerated, although no concrete proof exists of how they were suspended. However, the shape of their loop is more adequate in function for another category of object – as the small lid of an oval container to which it was attached by running a nail through the three-part jointed tubular channel, created by fitting the central joint as a composite part of the container, in the opening of the loop. A confirmation of such use is shown by an unpublished completely preserved box with an identical lid, but with a different figural relief image, from the collection of M. Ilkić, which comes from the Ilok region in Slavonia. The question remains unanswered as to whether these were seal-boxes or objects of a somewhat different purpose, but also in essence of a protective character.

It is known that Roman private and official documents were verified with seals, to the extent that, for example, military diplomas could bear as many as seven seals from witnesses to guarantee the accuracy of the transcription. The sensitivity of the wax meant that the stamped seals always had to be protected by some metal covering. Viewed in this capacity, the oval medallion-lid should be considered as an administrative requisite used in issuing official certificates. However, some of the images stamped on them hardly represent a neutral figural decoration, and in terms of content they must have been directly connected with their purpose. The victorious female warrior deity, even in an Amazonian version, suggests sealing a recognition that exceeds the level of sporting and other competitions, signalizing an important military achievement and results as the reason for awarding decorations marked by a military symbolism.

Part II

New Finds

The material presented in the second half of the volume consists of the small finds excavated in 2003 at the site of St. Quirinus (Fig. XXVI), in the framework of the most comprehensive multi-year systematic field research ever undertaken and carried out in *Siscia*, which is otherwise densely occupied by the urban architecture of the modern city. A city graveyard with a cemetery chapel dedicated to St. Quirinus Fig. II) was located in the 19th century at this site on the northern side but within the walls of the Roman nucleus.

In the excavation campaigns at this site, which will soon be turned into an archaeological park, to date architectural remains (public and residential) have been discovered, along with sections of streets, parts of walls with city gates, etc., which will be published in a separate volume.

This publication covers the small metal and glass finds, as well as certain other fragments that were found with the others. The material has gone through only the basic essential cleaning process, and the depictions on some of the bronze artifacts are difficult to recognize, in some cases almost invisible (Fig. 46).

The finds are classified according to the material from which they are made.

Fig. XXVI. View of the site of St. Quirinus.

Iron

The iron finds consist of: a double hook for hanging – no. 1 (Pl. I, 1, Fig. 1), pieces of thick circular and rectangular bars – no. 6 and no. 7 (Pl. I, 6, 7, Fig. 6, Fig. 7) – probably parts of tools,[1] spikes and nails of large dimensions – no. 8, no. 9, nos. 11-13, no. 19 (Pl. I, 8, 9, Pl. II, 11-13, 19, Fig. 8, Fig. 9, Fig. 11- Fig. 13, Fig. 19), suitable for attaching wooden roof beams and other elements of heavy structures, and nails of several standard sizes – nos. 14-18, nos. 20-22 (Pl. II, 14-18, 20-22, Fig. 14 – Fig. 18, Fig. 20 – Fig. 22), such as are usually used in woodworking.[2] The state in which no. 4 (Pl. I, 4, Fig. 4) was found does not permit even an approximate possibility of identifying the object of which it was a part. Item no. 5 (Pl. I, 5, Fig. 5) seems like an entire object, which in its original form might have represented a wedge with a hammered parallelogram-shaped upper section or a tool with a chisel-like flattened broader side, but it appears unsuitable in its present state for either of these purposes. Example no. 2 (Pl. I, 2, Fig. 2), is a pin with an unknown original length of the shank or size of the head,[3] while the shapeless handle no. 3 (Pl. I, 3, Fig. 3), appears to be an unfinished, discarded piece, which was probably already deformed when it was being made.[4] Due to the extremely poor state of preservation of item no. 10 (Pl. I, 10, Fig. 10), broken off directly beneath the short full neck, the original lines of its shape have been irreversibly lost. According to its present appearance, it is closest to the point of a catapult bolt with a pyramidal head and a hollow socket for the shaft, typical for the Roman army in the Republican period.[5] The identification should be taken with caution, as the *Siscian* example is missing the tip and the hollow socket with perforations for the nails, and also the original length of this piece, which is essential for typological classification, remains unknown.

The double hook, the handle, and the wedges and nails belong to forms that were produced throughout the entire Roman period and cannot be closely defined chronologically, except indirectly in intact closed units.[6]

The iron finds also include a piece of iron, no. 23 (Pl. II, 23, Fig. 23), which judging from the relatively regular outline of its exterior edge and the hollow in the middle probably represents a destroyed object or a part of it, as well as lumps of iron slag, no. 24 and no. 25 (Pl. II, 24, 25, Fig. 24, Fig. 25).

White metal

This category comprises half-finished products most probably made from lead, nos. 26 and 27 (Pl. III, 26, 27, Fig. 27), the latter possibly bars of solder, the amorphous pieces of lead nos. 28 and 29 (Pl. III, 28, 29, Fig. 28, Fig. 29), a fragment of sheet metal whose hardness excludes the possibility that it is made of tin and indicates it might be of silver, no. 30 (Pl. III, 30, Fig. 30), and the almost certainly silver ring, no. 31 (Pl. III, 31, Fig. 31), with the very simplest, chronologically unspecific form.[7]

Bronze

Among the bronze rings as well, the above simple neutral form that can extend chronologically throughout the entire Roman period was again present, no. 32 (Pl. III, 32, Fig. 32). Another example, no. 33 (Pl. III, 33, Fig. 33), was well made and originally must have had an insert of some other material in its circular socket, belonging to a solitary form without direct analogies that cannot be assigned to a specific date within the span of the 1st-3rd centuries.[8] The form of example no. 34 (Pl. III, 34, Fig. 34) should be considered a distant derivation of the snake-like rings from the end of the 2nd century and later,[9] when very gradually decoration with the appearance of notching began to be applied, what is known as chip-carving (Kerbschnitt). The distinctive appearance of the ring no. 35 (Pl. III, 35, Fig. 35), which gives the impression of a modest imitation of a more luxurious unique piece of jewellery, makes it difficult to find a direct analogy,[10] but a rosette with more than four sockets makes it closer to medieval or even later tastes, and would indicate a post-Roman artefact.

The wire earring no. 36 (Pl. III, 36, Fig. 36) in its present state does not offer the possibility of a more specific chronological classification in the framework of the Roman period. Its shabby appearance is hard to explain as a result of being worn from wearing and it was more likely an unsuccessful item discarded during working.

A moderately preserved Aucissa fibula, no. 37 (Pl. III, 37, Fig. 37), represents an inexpensive mass produced example

---

[1]   No. 6 and no. 7 could perhaps be parts of a mortise chisel-punch and chisel (BRONZI ANTICHI 2000: 212, no. 430; DA AQUILEIA AL DANUBIO 2001: 160, no. 108, b, c).

[2]   So far, no general typological framework for the numerically extensive scale of nail forms has been established, nor have they been classified according to purpose.

[3]   Assuming a longer shank and the presence of decoration under the corrosion layers in the centre the pin could be classified as a hair pin, or in the case of a smooth shank with the present length as a clothing pin.

[4]   Despite its massiveness, the handle is awkward, because the deeply bent arch meant it could be mounted only on a vessel with too small a spout diameter.

[5]   The origin of a bolt for a *pila catapultaria* with a pyramidal head and a long hollow socket has been placed at the end of the 3rd or in the first half of the 2nd centuries BC (BOŽIČ 1999: 73, Fig. 3, 2, 3, Fig. 4, 1; HORVAT 2002: 155, 164, 167, 185, Pl. 3, 1-4, Pl. 14, 7-9), but it was also still used in the 1st century AD (G. ULBERT 1969: 51, Pl. 46, 9, 16-18) and even later (WALKE 1965: 153, Pl. 108, 24).

[6]   Iron nails were also placed by the deceased in graves: well dated examples of the same form, 5.5 - 13 cm long, can be found at the Ptuj cemeteries in the context of finds from the 1st-4th centuries (POETOVIO II 1999: 19, 263, Pl. 3, 2, grave 9: end of the 1st – beg. of the 2nd cent.; 23, 266, Pl. 6, 3, 4, grave 20: second half of the 1st – beg. of the 2nd cent.; 205, 395, Pl. 135, 1, 2, grave 617: reign of Antoninus Pius – first half of the 3rd cent., etc.).

[7]   The form is timeless, appearing in jewellery from its very beginnings, and hence rings from the Greek classical period, for example, cannot be distinguished from those of the Roman period (BECATTI 1955, Pl. H, 1, 4).

[8]   It can be loosely tied to rings with ends fastened with wire coils from the 3rd century (HENKEL 1913: 228, 229, Pl. XXIX, 714, 721, 725)

[9]   Such rings almost always have a ribbed or incised hoop (HENKEL 1913, Pl. XXX, 746, 757).

[10]   It has certain parallels in rings with hollows in the central plate and the shoulders with insets of enamel from the middle Imperial period (HENKEL 1913, Fig. 240, Fig. 250)

of these two-part brooches, cast together with the final button. In terms of the lost or now unclear remaining details, which are insufficient for classification into variants,[11] the fibula can be placed in the context of the entire 1st century.

The cast pendant, no. 38 (Pl. III, 38, Fig. 38), made in one piece, has a rhomboid form with a punched surface decoration based on a floral motif, which is absent from Bishop's typology of the forms of pendants on Roman equestrian equipment (Bishop 1988). Pendant no. 38 represents a new, sixth example within one of two series (Siscia 1995, Pl. 29, 248-250) of identical rhomboid pendants from the same workshop from the 1st century that belonged to military equipment, which so far have been found in large amounts specifically in Sisak.[12]

The condition in which no. 39 was found (Pl. III, 39, Fig. 39) apparently would not allow consideration beyond the conclusion that it was a broken off part of an object that was badly cast and discarded in the manufacturing process. However, the remains of serration on the lower edge imply that it might perhaps represent the fan-shaped lower section of a large plate pendant or a *pelta*form openwork fitting, of the type Bishop 3c or 3d, which belonged to the equipment of the Roman cavalry in the 1st century. This conjecture is supported by such examples of a similar size (Siscia 1995, Pl. 28, 241), which were previously found in Sisak.[13]

Fitting no. 40 (Pl. III, 40, Fig. 40) can be classified among medium sized *phalerae*, probably from a chest strap or the scabbard of a sword, and dated to the 1st century. Originally it must have been a valuable piece coated, either partially or wholly, with a layer of a white alloy that in combination with the *niello* inserts, gives the impression of silver. A certain hesitancy in terms of the original appearance is caused by the remains of a poorly visible central decoration: it is unclear whether the inadequately sized hollows in the middle of the plate also had a *niello* or other fill or served for attaching additional relief decorations, which was considerably more common on similar mounts.[14]

The burnt mount fragments, no. 41 and no. 42 (Pl. III, 41, 42, Fig. 41, Fig. 42), are too small to attempt reconstruction of the type of openwork decorative motif, it is only possible to place it in the middle Imperial period on the basis of the workmanship.

For the example with a frontal appearance of a Bronze Age axe, no. 43 (Pl. III, 43, Fig. 43), this was most probably

the final section of a quite atypical Roman handle from a casserole or *trulla*,[15] which cannot be classified chronologically.

The globular buttons no. 44 and no. 45 (Pl. III, 44, 45, Fig. 44, Fig. 45) both in form and appearance suggest a later period beyond the boundaries of the Roman chronological framework, exhibiting similarities with such globular buttons on medieval male attire and horse equipment. They also deviate in terms of the manufacturing process: the workmanship of the first button is unclear because of damage, while for the remaining example it is evident that it was cast in one piece.[16]

For the three-lobed appliqué no. 46 (Pl. III, 46, Fig. 46) it is also difficult to establish a chronological framework, but it differs in appearance from Roman period items of this type. The depiction on it is in positive relief, and is particularly interesting but at the same time contradictory because of the arrangement of the profile into some kind of *gryllos* composition, such as are often engraved on ancient gemological material, which assumes knowledge of the glyptic production of the classical period. The fitting of the faces one into the other, as well as the design of the dolphin with its snout in a Greek version indicates a direct reliance on representations in the Hellenistic and classical manners. However, even viewed as some iconographic synthesis, the profiles cannot be tied to any known character, motif, or scene within the Roman repertory of images, nor can any concrete symbol or attribute be recognized in the floral decoration. These discrepancies, despite the classical connotations, would indicate that this was a post-Roman product.

The compact handle no. 47 (Pl. IV, 47, Fig. 47) with a mask of *Silenus*[17] displays certain autonomous characteristics, while analogies are evident only in individual composite decorative parts. According to the braided section, which actually represents joined twisted animal horns, and the twisted beard at the bottom of the mask like a goatee (connotation to the image of *Pan*), it belongs to a version of handles combined with animal masks. It is original and somewhat atypical as it joins the details of more modest and plain handles with elements of attachments from luxurious vessels,[18] attempting to achieve with a white coating the appearance of an object made from a more valuable metal. For a mass-produced item, the handle was cast in an above average quality, the somewhat exaggerated Silenus mask

[11] Related examples come from early Roman military sites (G. ULBERT 1969: 37, 38, Pl. 22, Pl. 23).
[12] Three earlier finds, in a variation with a central "button", have been previously published (KOŠČEVIĆ 1991: 52, 53, 129, Pl. XV, 227-229); while another two earlier finds, one with a button and one with a flat plate identical to no. 38, but somewhat larger, were more recently published, with citations of several analogous examples from western European sites (RADMAN LIVAJA 2004: 91, 92, 133, br. 250 – 254).
[13] Where they were also produced (KOŠČEVIĆ 1997: 46, Pl. 2, 55, 56).
[14] Among the circular fittings with a similar diameter from Sisak found to the present, the central decoration was frequently made separately and covered most of the plate (KOŠČEVIĆ 1991, Pl. XXXI, 413, 414, Pl. XXXIII, 453).

[15] Such simplest forms of handles appear very rarely in Pannonia (RADNOTI 1938, Pl. IV, 20, Pl. VI, 25, 26, Pl. XVI, 12, 14), but are somewhat more common at military sites (G. ULBERT 1959: 95, 106, Pl. 23, 11, Pl. 64, 1; WALKE 1965, Pl. 114, 3, 4).
[16] Globular buttons from Late Antiquity were cast in two parts and have a circular opening in the lower half, opposite the half with the loop for suspension (T. ULBERT 1981, Pl. 31, 4).
[17] Among handles with anthropomorphic masks, those of *Silenus* are most often combined with floral motifs (BOLLA 1994: 80, 81, Pl. LXXXIII, 88). A form based on horns is present on an incomplete handle from Magdalensberg with an anthropomorphic lower section with an entirely different style, dated to the second half of the 1st cent. BC (DEIMEL 1987: 135, Pl. XVI, 3).
[18] Often made of silver (KÜNZL 1984, Pl. 52, 2).

was naturalistically conceived, with an expressive plasticity. The choice of motifs and their combination reveals an effort aimed at highlighting a symbolism belonging to the *Dionysian thiasos*.[19] The closest parallels can be found in examples in the form of horns with animal protomes, of similar size, one of which comes from a grave dated to the Claudian period (Breščak 1982: 25, 55, Pl. 12, 117, 118). The handle belonged to a small jug probably used primarily for cult purposes, while chronologically it can be placed in the period of the 1st and 2nd centuries.

Neither hook no. 48 (Pl. IV, 48, Fig. 48), for hanging lamps, weights, or other items,[20] nor item no. 49 (Pl. IV, 49, Fig. 49), can be chronologically assigned to a narrower range than the 1st – 4th centuries, in which such practical objects were permanently utilized. For the latter item, because of the deformation of one arm and damage on the other, uncertainty evidently exists in terms of the purpose: did this piece have the function of a clamp or hoop for suspension after its arms had been driven into and firmly attached to a background surface, or was it in fact a pair of pincers (*volsella*), as the size and basic form can be the same for both objects. However, the slightly narrowed band of the arms as well as the preserved rounded end on one of them would define example no. 49 as a closed hook, probably for attaching a handle or hoop to a wooden base.

The bronze finds also included a certain number of fragments of sheet metal, rods, and bars, as well as pieces of burnt items and those destroyed by corrosion: nos. 50-55, nos. 62-64 (Pl. IV, 50-55, Fig. 62 - Fig. 64, Fig. 50 - Fig. 55, Fig. 55 a, Fig. 62 - Fig. 64).

The sewing needles no. 56 and no. 57 (Pl. IV, 56, 57, Fig. 56, Fig. 57), the first of which must have served to connect net-like or thicker and coarser material, also belong among everyday necessities with a continuity extending over all four centuries.[21] Both shapes of the eyes appear to be somewhat chronologically synchronous: examples with an oval eye from Magdalensberg belong to the early Imperial, i.e. the pre- and early-Augustan periods (Deimel 1987: 69, 70, 221-225, Pl. 49, Pl. 50, 1-5), while the pins with a rectangular eye from Aislingen were dated to the existence of the *castrum* itself, which is placed in the early Vespasian period (G. Ulbert 1959: 95, Pl. 24, 7). The pieces of wire with pointed ends, nos. 58-61 (Pl. IV, 58-61, Fig. 58-61), also represent parts of pins or needles, and the last could also be the pin from a fibula.

The probe no. 65 (Pl. IV, 65, Fig. 65) primarily represents

a medical device, specifically a variant with a bent *spatula* on one end and the other end pointed (*auriscalpium*, ear probe, Ohrsonde),[22] which was used both for medical and cosmetic purposes. The production of such probes has been confirmed throughout the entire Roman epoch and it is difficult to date them more specifically within the span of the 1st – 3rd centuries.

Bone

Four bone artefacts were found, of which the first, the entirely preserved no. 66 (Pl. V, 66, Fig. 66), was a probe identical in shape to the bronze example above (no. 65), but somewhat larger in size,[23] while the fragment no. 67 (Pl. V, 67, Fig. 67) could also have belonged to such an instrument, or it might have been the shaft of a hairpin.

Example no. 68 (Pl. V, 68, Fig. 68) could perhaps represent part of a small container for storing needles or other small implements.

The animal tooth no. 69 (Pl. V, 69) has been included because of the possibility that it originally could have had some element for suspension, in which case it could be classified among such pendants, which were quite popular in the Roman period. The incomplete preservation means that its identity as a pendant remains a conjecture.

Stone

The stone finds consisted of a whetstone, no. 72 (Pl. V, 72, Fig. 72), with an uncertain time frame, a probable mosaic cube, no. 71 (Pl. V, 71, Fig. 71), and tentatively part of a mould, no. 70 (Pl. V, 70, Fig. 70), or some other object from a more recent period.

Pottery

Of the pottery finds within the material discussed here,[24] three small fragments remained on the basis of which it is somewhat possible to determine the type and variant of the original vessel. Fragment no. 74 (Pl. V, 74, Fig. 74) belongs to a type of fine thin-walled table ware from the 1st and part of the 2nd centuries, and apparently was part of a beaker or small bowl.[25]

Fragment no. 75 (Pl. V, 75, Fig. 75) was broken off from the disc of an oil lamp: on the basis of the minimal remains of the decoration – one arcade section of an egg-shaped row – it appears possible to attribute it to the clay lamp type with a sharply angled nozzle and volutes, variant Loescke I or V/Ivanyi I or VI, from the first half of the 1st century (Vikić

---

[19] Vessels with handles featuring decorative elements containing *Dionysian* themes have previously been known from *Siscia*, dating from the 1st to the 3rd centuries (ANTIČKA BRONZA 1969: 127, 128, Fig. 227, Fig. 233).
[20] The diameter of the loop varied considerably in size, while the chains were mostly woven from doubly bent circlets and have the appearance of braids (DA AQUILEIA AL DANUBIO 2001: 108, Fig. 23).
[21] The bronze pins, 10.5, 14, and 19.2 cm long, discovered at the Ptuj cemeteries, mainly belonged to the 3rd and 4th centuries (POETOVIO II 1999: 41, 69, 153, 279, 298, Pl. 19, 1, Pl. 38, 7, Pl. 472, 14, graves 86, 177, 472).

[22] Several hundred examples of probes of this type come from *Siscia*, including bronze and bone examples (GREGL 1982: 179, 180, Pl. 3, 2 – 8).
[23] Such probes are also cited in the professional literature using other terms that characterize their purpose: *auricularium, specillum*.
[24] The pottery material from the 2003 excavations will be separately analysed and published.
[25] Related examples from Drenje are dated from the 1st to the middle of the 2nd centuries (DRENJE 1987: 33, 34).

Belančić 1971: 104, 133, Pl. IV, 2, Pl. XIX, 3; Neviodunum 1978: 40, 82 Pl. LVI, 1).[26]

Fragment no. 76 (Pl. V, 76, Fig. 76) is from a vessel that belongs to coarse pottery of a simple La Tène-like form derived from Celtic production, which can be chronologically assigned to the 1st century.[27]

Finds from other materials consist of a miniature piece of wall plaster, no. 73 (Pl. V, 73, Fig. 73), small bits of carbonized or dried wood, nos. 213 n-o (Fig. 213 n, nj, o), and the finds made of jet, no. 208, no. 214, and no. 215 (Pl. XIV, 208, 214, Fig. 208, 214, 215).

Glass

The greatest quantity, i.e. more than two thirds of the total small finds discovered in 2003 and presented here, consisted of glass artefacts. This large group, however, does not include a single vessel that could be entirely reconstructed. The group contains over a hundred fragments broken off the rims and necks of vessels, from their walls or their bases, as well as pieces of handles, while over two hundred tiny fragments represent shards of glass, including window glass, which can be discussed only statistically. The group also contains several pieces of jewellery, circular plates for games, mosaic cubes, and little pieces of raw glass.

In the analysis of the material found in the glass group, difficulties are created by the lack of characteristics for determining the types and variants, such as the total dimensions of the vessel and the full profile. This results in a limitation of insight into the numeric volume of forms as well as their purpose, as in a great number of cases it is not possible to identify even the type of vessel to which the fragment belonged. This lack of information elements is reflected in the chronological classification, which is mainly limited to typological dating.

The vessels possessing almost all basic elements for identification included *balsamaria* no. 77 and no. 78 (Pl. VI, 77, 78, Fig. 77, Fig. 78), of tubular form (Isings 8), which were widespread with numerous analogies,[28] and are approximately dated to the 1st and the first half of the 2nd centuries. The necks nos. 79 – 82 (Pl. VI, 79-82, Fig. 79 – Fig. 82) most likely belonged to *balsamaria* with an egg-shaped, pear-shaped, or globular body, which are chronologically congruent to the previous ones, although

analogous examples are quite scarce.[29] The situation is similar for examples nos. 83 – 86 (Pl. VI, 83-86, Fig. 83 – Fig. 86),[30] while in relation to the small bottle with an unusual curve to the neck, no. 87 (Pl. VI, 87, Fig. 87), as well as a fragment with a larger diameter rim, no. 89 (Pl. VI, 89), and a fragment (no. 88) with an urn shape, (Pl. VI, 88, Fig. 88), which most probably belonged to a square-bodied flask, the search for parallels with similar rim diameters remained without results.

The glass flask no. 90 (Pl. VI, 90, Fig. 90), considering the wall fragments, represents the best-preserved example, and at the same time it is one of only a few examples where fragments of the same vessel were found. Its neck and shoulders have all the characteristics of the later type of flask with a straight-cut rim and a globular receptacle (Isings 92), including the distinctive ribbed decoration.[31] It can be placed in the 3rd century, and it has almost direct parallels with a contemporaneous example made of green glass from *Salona* (Transparências 1998: 143, no. 100).

The neck no. 91 (Pl. VII, 91, Fig. 91) with a tri-lobed spout (Isings 56a) belonged to a jug with a curved handle above the rim, a globular body, and a ringed foot,[32] dated to the second half of the 1st and the 2nd centuries. Judging by the height and line of the neck, it is closer to the variant with a long neck and a flattened body, like one small jug attributed to a northern Italian workshop and dated to a span from the middle of the 1st into the 2nd century (Larese-Zerbinati 1998: 103, 125, no. 11, Pl. XIV, 11). Because of the incorrectly placed handle on no. 91, it should be considered a reject, discarded during manufacture.

A jug with a three-lobed rim could also be the source for fragment no. 126 (Pl. IX, 126, Fig. 126), and handle no. 184

---

[26] Large amounts of examples of this type are known from Sisak (VIKIĆ BELANČIĆ 1971: 134 ff.).

[27] At Drenje and Drnovo this type of pottery is placed chronologically in the second and third quarter of the 1st century and in the first half of the 1st century, respectively (DRENJE 1987: 27, 28, Pl. 1, 4, 8, 12; NEVIODUNUM 1978: 39, 78, Pl. XLV, 3, 5, 6, 10, 11).

[28] Close similarities to the Sisak specimens is displayed particularly by examples of northern Italic production from the early Flavian period, i.e. the second half of the 1st century (CASAGRANDE-CESELIN 2003: 115, 116, 118, no. 116, no. 123).

[29] Example no. 79 is related to *balsamaria* from the first half and middle of the 1st century of northern Italian production, and no. 80 to *balsamaria* from the end of the 1st – 2nd centuries, whose production has been confirmed in all parts of the Empire (CASAGRANDE-CESELIN 2003: 100, 101, 184, no. 71, no. 72), while no. 82 is similar to a *balsamarium* from a grave dated to the reign of Domitian – the beginning of the 2nd century (POETOVIO II 1999: 108, 326, Pl. 66, 1, grave 321).

[30] No. 84 has certain parallels to examples from the end of the 1st and the beginning of the 2nd century (BIAGGIO SIMONA 1991: 149, Pl. 25), no. 83 and no. 85 display the curved rim characteristic for *balsamaria* of the 2nd century and later periods (CALVI 1969, Pl. 21), while the rim of no. 86 has an appearance similar to that of *balsamaria* indirectly dated from the second half of the 1st to the beginning of the 2nd centuries (POETOVIO II 1999: 19, 35, 263, 274, Pl. 3, 3, Pl. 14, 8, graves 9 and 71).

[31] The type appeared in the 3rd century and became popular in the 4th century, examples varying primarily in the height of the neck and the shape of the spout, while in terms of production, along with eastern Mediterranean workshops, western production centres are also cited, along with presumed northern Italic or central European production (CALVI 1969, Pl. 23; ZAMPIERI 1998: 133, 135, no. 230; TONIOLO 2000: 113, no. 266; CASAGRANDE-CESELIN 2003: 83, 84, no. 52).

[32] Three-lobed jugs are documented in various parts of the Empire in the period from the middle of the 1st to the beginning of the 3rd centuries, with the greatest concentration from the end of the 1st to the first half of the 2nd centuries. They vary in terms of the neck and the receptacle, and are attributed to the workshops of northern and north-eastern Italy (TONIOLO 2000: 115, 116, no. 269; CASAGRANDE-CESELIN 2003: 26, 126, no. 149; LARESE 2004: 63, 64, Pl. XVII, 378, 380, LXXII, 11 ff.). No. 91 from Sisak is similar to jugs found at Magdalensberg (CZURDA RUTH 1979: 140-142, Pl. 8, 1049, 1051, Pl. 16)

(Pl. XIII, 184, Fig. 184), as well as the small rim fragment no. 186 (Pl. XIII, 186, Fig. 186).

Examples no. 92 and no. 93 (Pl. VII, 92, 93, Fig. 92, Fig. 93) with ground decoration[33] represent parts of beakers of colourless glass, for which it is difficult to find related analogous examples adequate both in terms of the form and the decoration. The first example is closest to a small cylindrical goblet with a ringed foot and a decoration of two rows of ground round facets in the upper part of the body, which was hypothetically attributed to an Alexandrian workshop and dated to the 1st century, with certain reservations about this considering the lack of analogies (Ravagnan 1994: 177, 182, no. 357). For the second example, the possibilities of comparison are primarily lacking in reference to the workmanship of the decoration. The shape of the decoration itself in the form of stretched perpendicular ovals is analogous to the so-called arcade motif (arched ribs or Bogenrippendekor),[34] which was carried out either in a technique of moulding the walls so that the ribbed ovals overtop the walls or in a technique of applying molten threads to the walls, but always as a relief decoration and not as a decoration sunk into the walls with a facet-cut edging like on the example from Sisak. Both examples decorated by grinding, no. 92 and no. 93, would correspond chronologically to the second half of the 1st and the 2nd centuries.

Example no. 94 (Pl. VII, 94, Fig. 94) is the stem of a goblet that in terms of appearance and size corresponds to Late Roman-Early Medieval forms of chalices (Isings 111).[35] It has a direct parallel to an example attributed to a northern Italian workshop and placed chronologically in the period of the 6th-7th centuries (Casagrande-Ceselin 2003: 92, no. 61), where the piece from Sisak should also be indirectly assigned.

The examples nos. 95 – 103 (Pl. VII, 95-103, Fig. 95 – Fig. 103) encompass fragments of monochromic ribbed bowls (Isings 3),[36] of which the first three rim fragments, no. 95 – no. 97, appear close to another variant of this shape (Isings 3b), and exhibit similarities with some bowls of northern Italian provenience, dated to the 1st century (Ravagnan 1994: 177, 178, no. 345; Zampieri 1998: 160, no. 256, no. 257), like those from Magdalensberg (Czurda Ruth 1979: 26 – 33, Pl. 1, 83, 114, 278, Pl. 12). The bowl to which fragment no. 96 belonged – the only one made of dark blue glass – should perhaps be considered the earliest example within the chronological span from the second half of the 1st century to the beginning of the 2nd century, in which all nine fragments can be placed.

The type of vessel to which fragment no. 140 belonged (Pl. IX, 140, Fig. 140), with differently formed vertical ribs,[37] is difficult to determine despite the size of the preserved piece.

The fragment no. 104 (Pl. VIII, 104, Fig. 104) composed part of a dish with a flat base, perpendicular walls, and a simple slanted rim of polychromic glass made in the *millefiori* technique,[38] which can be placed in the 1st century. Of similar known examples found in Italy and Dalmatia (Zampieri 1998: 163, no. 266, no. 267; Transparências 1998: 89, 184, 185, nos. 173-175), the closest to the dish from Sisak is a piece from Zadar dated to the end of the 1st century BC – first half of the 1st century AD (Ravagnan 1994: 223, 225, no. 453).

The tiny wall fragment no. 137 (Pl. IX, 137, Fig. 137) was also made of polychromic glass, in a more modest combination of two colours.

Judging from the line of the cross-section and the approximate diameter, examples no. 105 and no. 106 (Pl. VIII, 105, 106, Fig. 105, Fig. 106), must also have belonged to small plates or platters.

Fragments no. 162 (Pl. XI, 162, Fig. 162)[39] and no. 178 (Pl. XII, 178, Fig. 178)[40] were probably parts of beakers or small flasks with thick walls.

The tubular fragment no. 141 (Pl. IX, 141, Fig. 141) could have belonged to a funnel (Isings 74) or a device for testing

---

[33] The technique of faceting by grinding appeared on products from the 1st and 2nd centuries, and a ground decoration of facetted rhomboids is common on bell-shaped beakers from the middle of the 1st to the middle of the 2nd centuries (TRANSPARÊNCIAS 1998: 88, 89, 174, 193, nos. 152, 153, 192; POETOVIO I 1999: 76, Fig. 63).

[34] Ribbed decoration with the appearance of arcades can most often be seen on the bell-shaped beakers of type Isings 33 (Arkadenbecher), attributed to northern Italic workshops active in the period of the second half of the 1st – middle of the 2nd centuries (RAVAGNAN 1994: 126, 131, no. 248, no. 249; ZAMPIERI 1998: 132, 134, no. 227; CASAGRANDE-CESELIN 2003: 20, 62, no. 11).

[35] A goblet with such a stem from Verona belongs to a Late Roman context of the 5th-6th centuries with the presumed appearance of the form as early as the 4th century, while similar examples are also known from the Lombard cemetery near Vicenza (CASAGRANDE-CESELIN 2003: 27, 91, 194, no. 60, 307).

[36] Ribbed bowls were widespread and produced throughout the Empire in the period from the end of the 1st century BC to the first half of the 1st century AD. Monochromatic bowls were replaced by bowls made of mosaic glass (Rippenschale), while greenish-blue examples were particularly popular in Italy, the western provinces, the trans-Alpine region, and in Dalmatia and Pannonia, where they continued to be produced in the second half of the 1st century AD (LARESE 2004: 15, 16).

[37] In terms of shape, with ribs narrow above and widened below, it is related to the globular flasks of type Isings 26b from the 1st century, while the line of the walls is similar to that of some goblets with smooth walls and a base with an indented centre or with a ringed foot of small diameter covered by a deeply lowered body, which are considered eastern or northern Italian products from the second half of the 1st – 2nd cent. (RAVAGNAN 1994: 177, 186, no. 365; CASAGRANDE-CESELIN 2003: 165, no. 240).

[38] Items made of mosaic glass, which was developed in the Hellenistic period, are mostly attributed to Alexandrian production and are dated to the period from the end of the 1st century BC to the middle of the 1st century AD. On the basis of the concentration of finds of a somewhat different shape and decoration from northern Italy, the Italic Roman period production of this glass had already been suggested (ZAMPIERI 1998: 158, 159), reinforced by more recent finds of debris from casting in Aquileia, where one of the production centres is hypothetically placed, and which could be the source for the finds from Slovenia and Croatia (LARESE 2004: 14, n. 14).

[39] Late Roman beakers (Isings 106) from the 4th century have such flat bases (CALVI 1969, Pl. 26, 2, 3).

[40] An equally highly indented base with the same diameter is present on a shorter cylindrical flask with a different line of the body from the second half of the 1st – 2nd centuries (TONIOLO 2000: 104, 110, no. 255).

wine (Isings 76),[41] fragments no. 143 and no. 144 (Pl. X, 143, 144, Fig. 143, Fig. 144) to a small container of the *olla* type or a small bottle, the bases no. 154 and no. 157 (Pl. X, 154, 157, Fig. 154, Fig. 157) probably belonged to beakers, while no. 153 (Pl. X, 153, Fig. 153) certainly belonged to a beaker or small bowl from the 1st century (Larese-Zerbinati 1998: 139, no. 49, no. 50; Casagrande-Ceselin 2003: 199, no. 316).

Within the category of pieces broke off from thickened bases, fragments nos. 172-174 (Pl. XI, 172, Pl. XII, 173, 174, Fig. 172 – Fig. 174) with circular and square relief ribs on the lower side represent parts of square bottles with two handles (Isings 90), which generally belong to the 1st and 2nd centuries, and also to later periods. Handles no. 185 and no. 189 (Pl. XIII, 185, 189, Fig. 185, Fig. 189) were also most probably broken off from bottles with a square base, while example no. 181 (Pl. XIII, 181, Fig. 181) probably belonged to a monochromatic cobalt blue small jug.[42]

Further involvement in deciphering the original vessel types of the rim fragments (Pl. VIII, 107-123 and Pl. IX, 124-131), the wall fragments (Pl. IX, 132-139), and the base fragments (Pl. X, 145-153, 155, 156, 158, 159, Pl. XI, 160, 161, 163-171, Pl. XII, 175-177, 179, 180), would be reduced to speculation.

The group of other kinds of glass objects includes what are probably mosaic cubes, nos. 194-198 (Pl. XIV, 194-198, Fig. 194 – Fig. 198), one pair of raw glass, with an appearance identical to that of fragment no. 213e (Fig. 213e).

All three game tokens nos. 199-201 (Pl. XIV, 199 – 201, Fig. 199 – Fig. 201) represent the same simplest form of *calculus*,[43] because of which they can be classified mostly to the 1st century and partly to the next.

The globular and hexagonal beads no. 202 and no. 203 (Pl. XIV, 202, 203, Fig. 202, Fig. 203), were among popular forms from the Middle and Late Imperial period,[44] while the so-called "melon" (Melonen) beads no. 204 and no. 205 (Pl. XIV, 204, 205, Fig. 204, Fig. 205), as one of the long-lasting forms most persistent in terms of lack

of change,[45] including their turquoise colour with its symbolism, cannot be chronologically determined well within their broad framework of the 1st-4th centuries and later.

The bracelet parts no. 206 and no. 207 (Pl. XIV, 206, 207, Fig. 206, Fig. 207) represent simple variants of this type of jewellery with a closed smooth circlet, placed approximately in the 3rd and 4th centuries. Italic finds of glass bracelets are attributed to eastern workshops – hypothetically those of Roman Egypt – and are placed in the lengthy period of the 1st – 4th centuries.[46] In Pannonia, where monochromic examples predominate while fragments with polychromic decoration are much more rare, glass bracelets appear later and are dated to the 2nd – 4th centuries, and it is generally accepted that they were also produced in that province.[47] Several examples of such bracelets are known from Sisak.[48]

The remaining finds consist of large and small pieces of raw glass, nos. 209-213, 213a-e (Pl. XIV, 209-213, Fig. 209 – Fig. 213, Fig. 213 a-e), a teardrop shaped fragment of glass br. 213 f (Fig. 213 f), several pieces of glass slag no. 213 g-m (Fig. 213 g-m), and over 200 shapeless fragments of various vessels.[49]

Jet

Fragment no. 208 (Pl. XIV, 208, Fig. 208) represents part of a bracelet with a closed circlet, made of jet and decorated in the characteristic manner for this material, with deep incisions, which can be chronologically assigned to the period of Late Antiquity.

Examples no. 214 and no. 215 (Pl. XIV, 214, Fig. 214, Fig. 215) are shapeless pieces of jet,[50] which represent the most interesting discovery among the material found in the 2003 excavations. The rare sporadically discovered

---

[41] Both devices are dated to the second half of the 1st century (CALVI 1969, Pl. 18, 1; RAVAGNAN 1994: 201, 234, no. 397, no. 398, no. 470).

[42] They are dated to the 1st century and attributed both to eastern (hypothetically Syrian-Palestinian) and western (Italic) workshops (CALVI 1969, Pl. 7, 2; RAVAGNAN 1994: 165, 166, no. 324; TONIOLO 2000: 160, nos. 268-270)

[43] In addition to the most common black and white, they also appear in green and blue colours, with a diameter of ca. 1.5 – ca, 3 cm. They were widespread and produced throughout the entire Roman world, and have been discovered in dwellings, graves, public areas, and particularly in structures with a military character. They appeared as early as the 3rd century BC on Italic territory, with the greatest diffusion achieved from the Augustan period to the 2nd century AD, while they were very rare in the 3rd – 4th centuries (LARESE-ZERBINATI 1998: 86, 88-91, nos. 158-166; CASAGRANDE-CESELIN 2003: 95, 96, 135, 136, no. 173, no. 175).

[44] Italic examples of globular beads are placed in the 3rd – 4th centuries, and polygonal beads in the 4th century (CASAGRANDE-CESELIN 2003: 39, 127, 130, no. 150, 157).

[45] They appeared in Europe in the pre-Roman period, achieving the greatest distribution in the first two centuries AD, and then fell into relative disuse; their popularity returned in the 4th century, and their production continued well into the Middle Ages (RAVAGNAN 1994: 173, 174, no. 337; CASAGRANDE-CESELIN 2003: 40, 66, no. 24, no. 25).

[46] The hoops of the bracelets were most often blue, green, and purple, additionally decorated with patterns of other colours; purely monochromic examples are very rare, while the occasional examples made of black glass were deeply ribbed (RAVAGNAN 1994: 159-161, nos. 313-318)

[47] The Pannonian examples made from cobalt blue glass are dated to the 2nd century, while bracelets of black glass predominated in the 3rd and 4th centuries (INTERCISA I 1976, Pl. 27, grave 19a), and remained popular later, surviving beyond the Roman period.

[48] This is true not merely of black monochromic examples, but also black bracelets decorated with polychromic spots of glass. Glass rings were also found with them, including discarded pieces that would support the hypothesis that glass jewellery was produced in Siscia (KOŠČEVIĆ 1996: 82-84, Pl.1, 2-6, 17).

[49] These fragments are marked on the photographs with the catalogue number of the individual item with which they were found with the addition of the letter "a", while the individually discovered fragments are combined on Fig. 216.

[50] "Jet is a variety of petrified coal, with a highly glossy black colour, and a hardness and workability suitable for processing. It is found in the form of pieces embedded in bituminous rocks, and it contains 85-95% bitumen. Under the microscope, the original wood can be perceived, whose tissues have been subjected to great pressure, i.e. deformed. It is considered to have originated from wood immersed in bituminous, then

finished jet products are casually and barely worked, while the find of raw material for carving or pieces left over after working, like these from Sisak, can be considered a true rarity of major importance. Jet, also known as "black amber" (both substances are resins), to which healing and protective properties were attributed, was used for jewellery in the European region as early as the La Tène period.[51] It was suitable for carving hairpins, pendants, beads, and miniature cases. Jet jewellery from Pannonia,[52] which mostly belongs to the period of Late Antiquity, is extremely rare. The price and fragility of jet caused it to be replaced by mass-produced, less expensive jewellery made from black glass. These are particularly well-represented by black glass bracelets whose outer circlet is divided by deep incisions, which faithfully imitate the specific decoration of jet pieces. Such glass surrogate examples are not yet known from *Siscia.*[53]

Some of the artefacts discovered during the excavations in 2003 belong to chronologically indistinguishable categories. Of the finds that could be dated more reliably, most belonged to the span of the first two centuries, while for some of them the chronological framework extended to into the 5th and 6th centuries, and even into the period after that. The finds encompass objects of everyday use: the inevitable nails, pins, handles, hooks, fragments of decorative mounts, fragments of glass vessels, and probably parts of other utilitarian items, as well as medicinal equipment, and some jewellery, mostly rings, beads, and bracelets.

The objects with military characteristics, along with pieces of weaponry in the form of arrowheads (Fig. 10), include pieces of military equipment – a rhomboid pendant (Fig. 38) and circular mount (Fig. 40): the latter could belong among *dona militaria*, and quite hypothetically this could also be applied to a silver ring of diameter appropriate for a man (Fig. 31), although it lacks an inscription. An Aucissa fibula can also be related to the military (Fig. 37), as well as – considering the number of fragments – parts of a glass ribbed bowl (Fig. 95 – Fig. 103) as a practical object of mass use, and gaming chips (Fig. 199 – Fig. 201).

The finds that testify to the production of metal objects – iron, bronze, and lead: iron slag (Fig. 24, Fig. 25), amorphous lead (Fig. 28, Fig. 29), pieces of fire-damaged or decayed bronze (Fig. 54, Fig. 55, Fig. 55a), sections of

destroyed objects (Fig. 4, Fig. 23), discarded half-finished products (Fig. 3), production waste (Fig. 50-Fig. 53, Fig. 64), also include a discarded piece (Fig. 39) that can be considered part of a mount or pendant of military affiliation, which suggests at least a minimal possible participation of the military in such production.

Along with the direct evidence for *Siscian* glass production – raw glass (Fig. 209 – Fig. 213, Fig. 213 a-e), remains of glass slag and cinders (Fig. 213 g-m), as well as a possible drop-shaped remnant from a blowpipe (Fig. 213 f) – it would be logical to expect that this could indirectly also include sections of vessels and numerous fragments. However, viewed as a whole, the large group of pieces and fragments of glass do not fully fulfil such expectations and can be considered in two ways.

Among the types of vessels that could be considered more comprehensively, the usual forms were noted in the group of *balsamaria*, flasks, jugs, ribbed cups, and plates, predominantly from the first two centuries. The presence of individual or rare forms, such as some examples of *balsamaria* and beakers (Fig. 79 – Fig. 86, Fig. 92, Fig. 93) and fragments of similar vessels (Fig. 139, Fig. 140, Fig. 162, Fig. 178), and discarded deformed pieces, such as a jug and part of a wall with attached glass (Fig. 91, Fig. 142), would signal their production at the site of discovery. In fact, even in conditions with a complete absence of material evidence, for such a strong regional production and distribution centre as *Siscia*, it would be possible to establish that it manufactured not merely simple, widely distributed glass products, such as the ubiquitous *balsamaria*[54] and the popular ribbed bowls, but also – on the basis of comparisons with metal products of superb workmanship[55] – the more demanding and sophisticatedly decorated items, such as engraved beakers and polychromatic dishes.

As the situation is actually the opposite and numerous arguments are available, all of the glass finds should be able to be included among the present material evidence for *Siscian* glass production. In fact, the better preserved sections of vessels are inseparable from the remaining "piles" of fragments, including those from window glass. The fragments, evidently, themselves mostly represent individual isolated pieces whose other component parts, other than in the case of flask Fig. 90, would be futile and pointless to seek, as they must have been piled up several times in order to end up in their present state. The possibility

---

sapropelic deposits, taking from them certain gel and oil substances that impregnated the wood, resulting in hardness and resilience. Its name in Croatian (*gagat*) comes from the name of the Gages River in Lycia in Asia Minor. The most well-known deposits of jet are in Yorkshire in England, where an entire local industry exists with beginnings in the pre-Roman period, and rich deposits also exist in Spain". The analysis of the samples was undertaken and the cited data were made available by Dr. Josip Velić, tenured professor at the Department of Geology and Geological Engineering, the Faculty of Mining, Geology, and Petroleum Engineering of the University of Zagreb, whom I sincerely thank.

[51] Along with jet, sapropelic coal from the northern Czech Republic was also utilized (MOOSLEITNER 1980: 95).

[52] Jet was imported to Pannonia from Britannia and Gaul from the beginning of the 3rd century (INTERCISA II 1957: 405, 418).

[53] In contrast, for example, to Drnovo, where such glass bracelets appear in addition to carved lignite examples (NEVIODUNUM 1978: 63, 68, Pl. XIII, 40, 41, Pl. XXIII, 21).

[54] The numerical predominance, related primarily to their function as offerings in the burial ritual, has been confirmed at the *Siscian* necropolises, where they represent the most common grave goods (NENADIĆ 1987: 85 – 93; WIEWEGH 2002: 10, 11).

[55] Metal production, particularly of bronze, had complex phases of manufacture and was to a certain extent more demanding than glass-working (ŠAŠEL 1974: 726, 727; KOŠČEVIĆ 1997). In terms of the manufacturing process, the latter is more complicated than the basic production of clay-based items, proven at Siscia for certain types of products: bricks, oil lamps, ceramic reliefs (NENADIĆ 1987: 97; VIKIĆ BELANČIĆ 1976: 86; ŠEPER 1954: 312), and naturally also than the lead production at Siscia (BAUER 1936: 33), and it could have been stimulated and justified only by a corresponding demand, which certainly must have existed.

exists that the state in which they were found was a result of collapsing, because of which the vessels were broken, but this implies that the parts remained together, i.e. that several fragments of the same vessel would be found in the same place. Given that this case is totally the opposite, it would be difficult to accept this hypothesis. It is evident, in fact, that the great majority of fragments were deliberately fragmented to occupy the smallest possible space, and should be treated as scrap glass, which could only have been intended to be re-melted, which is indirectly confirmed by pieces of raw glass, to which discards were usually added. However, defining its purpose as a store of half-finished material does not get us any closer to an answer as to the place where they had been produced. In fact, glass for re-melting could have been gathered and stored at the site of discovery,[56] but it could also have arrived in *Siscia* from any nearby production centre.

Although the *Siscian* glass producers certainly manufactured the majority of the types and forms represented in this find, the designation of the group of glass as half-finished raw material for further production takes away its credibility as proof for an absolute attribution to this manufacturer. Similarly, the similarity of form with the Italic finds from the period of the first two centuries when the trade connections – confirmed through the import of other types of items – with the northern Italian region were most lively as yet does not offer a fully objective image that would be a sufficiently strong argument for attributing the waste glass to the workshops there. In and of itself, the group of glass at this moment cannot be tied to any concrete workshop and its origin remains uncertain. In order for the Sisak find to be connected more firmly to production there, it would be necessary to examine thoroughly, verify closely, and study in detail the types and forms of vessels in the previously uncovered glass inventories stored in the collections of the museums in Sisak and Zagreb.

In addition, for a final resolution of the existing dilemma, which also incorporates the problem of the too lengthy chronological span for the creation of the glass vessels that extends over several centuries, it will be necessary to await the results of further excavations at the site of St. Quirinus.

The present general value of the group of glass finds lies in the fact that it indicates on the one hand quite major glass production in *Siscia*, whether these were merely production discards for further transport or there was a need for their acquisition, and on the other hand the characteristics of the workshop in which they were created.

No matter where it was located, the anonymous workshop

that this waste glass belonged to utilized all techniques of workmanship and almost all manners of decoration known in antiquity in production: moulding and blowing into a mould, free-blowing, forming tubular and full rims and ring-shaped feet by bending to the outer or inner side, with numerous variants in the degree of the angle (Pl. VIII, 106 ff, Pl. X, 145 ff); the colour, quality, and transparency of the mass and the presence of colourless (decoloured) glass, at times reduced, at its finest, to the thinness of foil (Pl. IX, 135) correspond to the high standards in antiquity, as do the decorative processes, which in addition to engraving and polishing facets on a lathe (Pl. VII, 92, 93), also included stamping (Pl. IX, 133), placing threads of other colours in the surface layer of the walls (Pl. IX, 128, 134) or in the glass mass (Pl. XIII, 181), applying molten bands of the same colour onto the walls of the vessel (Pl. IX, 139), and decorative mosaic glass (Pl. VIII, 104, Pl. IX, 137), which required the greatest technological level of workmanship. A large number of pieces of raw glass, so-called glass ingots – previously known from the *Siscia* region – would suggest the possibility that *Siscia*, in addition to glass goods, also produced the raw material for their manufacture (*officinae primariae*),[57] but the lack of traces in this find of the other necessary components for the production of raw glass would at present disprove such an assumption.

In a brief recapitulation of the results offered by the material discovered in 2003, the main conclusions are summed up. The find encompasses a chronological framework as defined by the earliest (no. 37) and latest (no. 94) examples, covering the entire period of antiquity and entering into the early medieval period, along with several specimens from more recent times, designated as post-antiquity.

In terms of content, this is a source of fresh information about several types of metal[58] and glass[59] production. Earlier findings about this last supplement are to date conditional knowledge about its production capacities via a group that would most probably nonetheless need to be defined as recycling glass. If it were to be established with certainty – through additional study of the comparative material – that it was produced from *Siscian* manufacture, this would completely correspond to expectations about the production possibilities of a city of colonial rank.

Numerous already published facts are known about

---

[56] Along with trade in finished products and raw glass, glass waste (cullet) was also traded, which was gathered in an organized fashion for this purpose. Several such prepared cargoes are known, like the find from the last quarter of the 1st century at Augsburg that contained fragments from more than 600 vessels stored in pits in the soil (ROTTLOFF 2000: 134, Fig. 107) or the crushed glass fragments placed in a barrel that composed part of the cargo of a Roman shipwreck discovered near Grado, thought to have been intended for Aquileia (TONIOLO 2000: 11).

[57] The production of raw glass, as a primary process, was separate from the secondary production of vessels, and it can be proven through the remains of lime and traces of quartz sand, which along with soda compose the basic components of glass. An example of trading in raw glass in large quantities is probably represented by the find of a sunken ship with such a cargo near the island of Mljet (FADIĆ 2002: 387).

[58] This has already been confirmed many times by the numerous finds that cover activities ranging from the metallurgical, with the nearby ore deposits in Ljubija and along the Japra and Sana Rivers, to the manufacture of many artistic-craft and other objects of everyday use from various metals and alloys.

[59] Glass production was also previously proven through unpublished finds of primary importance, including large fragments of smelting pots coated inside with a glass slag like a glaze and pieces of raw glass stored in the Archaeological Museum in Zagreb, as well as the discovery of Roman glass furnaces in Sisak (BURKOWSKY 1999: 67; VIDOŠEVIĆ 2003: 12, 13, Fig. 2).

the metal production of *Siscia*, mostly in bronze and lead, including data about the possible location where it took place – in one part of the banks of the Kupa River, hence within the Roman city but outside the walls. The river bed in this section, from which objects have been dredged for more than a century – including a considerable percentage of half-finished items – which today remains an inexhaustible archaeological source, is the origin of most of the *Siscia* finds. Thanks mainly to this unique source, *Siscia* today is the quantitatively strongest Pannonian site for fibula material, and the second in terms of the numbers of "perforated" bronzes.

In analysis of the *Siscian* finds, two characteristics can always be noted first: their mostly military nature and the Celtic inventiveness present in their formation.

The Celtic element cannot be specifically distinguished, but it is more or less reflected in several residual aspects, such as the modelling of physiognomy (pottery sculpture with intriguing stylistic characteristics: Siscia 1995, Pl. 11, 60-65; empty-handed *Mercury*: Fig. VII), as well as in the specific styling of decorative details (*lunular* pendants: Fig. XXIV), the design of the apertures on the perforated products (pendants: Fig. XIX, a, c, d; pierced bronzes: Siscia 1995: 35, Fig. 25), etc., as distant but not completely faded reflections of ancient Celtic influences and Celtic craft empiricism.

Although so far no secure traces have been found of the remains of military structures that would offer a marker for locating the zone of the military camp in *Siscia*,[60] the older finds, and to a certain extent these new ones clearly manifest themselves as military or at least to a certain extent contain connotations of a military context, particularly in relation to the manufacturing sector, returning once again to the fact that the city was in fact based on military roots.

The objects retrieved from the Kupa River have military characteristics, not merely in relation to the three most frequent types of fibulae: the widely distributed Aucissa and bulbous fibulae, along with the Pannonian multiple articulated type. A distinct military nature is exhibited by buckles and mounts from a *cingulum* (Siscia 1995, Pl. 25, 204, 205, 210), rivets from the belts of so-called military aprons (Siscia 1995, Pl. 30),[61] Roman cavalry harness pendants – from simple leaf-shaped to pendants with complicated decoration (Siscia 1995, Pl. 27-29, 248-250).[62] Some of the seal boxes are also related to a

military milieu,[63] and the images on some pendants have certain military or at least competitive characteristics (Fig. XV; Siscia 1995, Pl. 53, 568-570). A military purpose is particularly manifested on parts of the small *utilitaria* for seal protection with an emphasized belligerent and victorious official state symbolism embedded into their decoration (Fig. XXV; Siscia 1995, Pl. 46, 487-490).

The extensive production capacities of the facilities along the Kupa River can be perceived not merely in the discovered quantity of individual types,[64] but also in the proven simplification of manufacturing procedures in order to accelerate and increase productivity (Siscia 1995: 33, Fig. 24). All these recapitulated facts speak in favour of a primarily camp production of long duration that began with its foundation, probably as early as the end of the 1st century BC.

The most intriguing question that arose from reviewing the archaeological material from 2003 was related to the character of the newly uncovered production and the relations between the two production locations. The discovery of a production site within the city itself is not surprising and could be expected.[65] Similar manufacturing *insulae*, grouped for the production of items of the same type, must have existed, and it first seemed that this production centre at St. Quirinus was related to glass-working.

It turned out, however, that it was mixed, and in addition to glass working also included bronze, possibly iron, and jet craft activities, and surprisingly the presence of several objects of military character (Fig. 38, Fig. 40), which must have been created at the site of discovery. This presumes that production was not merely for the civilian but also for the military sector, and that they were not strictly separated.

But the actual production relationships between the St. Quirinus and Kupa sites, the motivations for proportions in the quantity and types of productions, and the conditions under which they were agreed are too tangled an issue, requiring many more valid arguments and evidence to be unravelled.

---

pure types of leaf-shaped, bird-shaped, and pierced pendants with a design based on a motif of heraldically opposed animal snouts (KOŠČEVIĆ 1991, Pl. XIII, 199 - Pl. XV, 219).
[63] Some 90 examples of seal boxes of various types come from Sisak, the majority manufactured there (KOŠČEVIĆ 1997: 46, nos. 88-98). This includes a series of around 20 examples with a meticulous stamped linear decoration with niello, which are extremely rarely found elsewhere: one example was found at Vindonissa, and one fragment at Ilidža near Sarajevo, both in the context of finds from the area of military camps (KOŠČEVIĆ 1991/1: 26, Pl. I, 15-20).
[64] For example, about 450 examples of fibulae of the same type, around 100 examples of pierced fittings and objects of similar purpose, etc.
[65] A production site for glass-working was proven for the city with the discovery of a furnace (Fig. II, 13), but probably also existed within the hypothesized trade-craft structure (Fig. II, 16), where it was apparently related to pottery production: a small group of fragments of various glass vessels also contained a piece of a ribbed bowl of blue glass, the unsuccessful rim section of some receptacle, and "the waste part of the glass with remains of a brick-like coating" (PREGLED 2000: 8, 21); it is interesting that fibulae of military types come from the same site.

[60] Such a possibility is indicated below the horreum and part of the walls with a tower (Fig. II, 14), where the foundations were discovered of an early Roman structure that was hypothesized to have supported a wooden superstructure – perhaps with the function of a military barracks, while among the discovered pottery material what are known as military plates were strikingly predominant (PREGLED 2000: 34, 35, 39).
[61] Of the so far known more than 280 pieces of such rivets, which at some other European sites were produced in legionary camp workshops, some 30 examples are from Sisak (KOŠČEVIĆ 1991: 80, 81, Pl. XXXIII).
[62] The numerous half-finished products of this group consisting of 23 crude casts of a simplified leaf-shaped form intended for further working (KOŠČEVIĆ 1997: 46) indicate a great number of metal decorations intended for the harnesses of the cavalry. In their final form, these are

Glass and metal production are related, as both involve thermal treatment, i.e the use of fire. They were joined at St. Quirinus by the manufacture of objects made of jet, which took place under the same roof. Even though the expression "raw material" has been used here in the absence of any more suitable term, it is uncertain how the raw jet would originally have looked, with the exception of the size of the pieces themselves, which must have corresponded to the largest dimensions of a bracelet for an adult individual. The deformed appearance of the Sisak examples of jet fragments – with one or both surfaces covered by hollows and protrusions with soft edges – bears witness to changes caused by high temperatures, which cannot be connected to the process of manufacturing in this material, as it requires only treatment in its natural state. In the case of the metal and glass finds, the damages to some of them could have been caused by chance exposure to fire and yet be interpreted as evidence of production. However, the resinous surface of the Sisak jet, created by lengthy exposure to fire to the extent that the exterior surface of the material began to melt, would rather indicate that the structure in question had been destroyed by fire. However, the reason for the destruction represents a separate theme, and this is only touched on by the observations expressed here, based on the phenomena exhibited by some of the finds. The question of how a manufacturing facility with a mixed craft production system, as discovered in 2003, could have been destroyed, is a more complex issue laden by a series of factors that require more extensive consideration.

The most significant result from the 2003 find is nonetheless the discovery of the jet, although it is not yet possible to indicate from what region it was acquired. It reached us in pieces with the appearance of debris remaining after manufacture or as the broken fragments of some larger pieces of raw material exposed to fire. No matter which phase of the production process they belonged, they still represent authentic proof of the production of jewellery from jet, which had not yet been documented for Pannonia. Thanks to these small fragments, *Siscia* has been added to the Roman centres where the production of jet jewellery has been definitely confirmed, and a new item has been added to the previously known assortment of goods that *Siscia* produced.

CATALOGUE

1. A double hook composed of thick rods bent into a loop, with the ends curved to the opposite sides. It is badly corroded, with a cracked surface from which layers peel off. The final section of the right hook is broken off. Iron. Dimensions: 52 x 72 mm. Along with this piece were found no. 23, no. 25, and no. 213 g. (Pl. I, 1, Fig. 1).

2. A pin (hairpin?) with an irregular head and an unclear original form of the shank. Badly damaged, the central and lower part are covered with a corroded layer. Iron. Length: 59 mm. It was found together with no. 5 (Pl. I, 2, Fig. 2).

3. A handle of irregular form and coarse workmanship. It is corroded and badly damaged in places with a cracked surface, the arch is broken, and the end of the left section is missing. Iron. Dimensions: 67.5 x 55 mm. It was found together with no. 20 and br. 30 (Pl. I, 3, Fig. 3).

4. An exteriorly misshapen piece of metal under a thick corrosion layer retaining small pieces of brick and coal. The lengthwise break in cross section shows a visible part of the item consisting of bent thick shiny sheet metal. In the lower section, where the end was broken off, the sheet metal was bent to the side, i.e. folded. Iron. Dimensions: 33.5 x 21 x 15 mm. It was found together with no. 14 and br. 50 (Pl. I, 4, Fig. 4).

5. A spike (?) of square section with a flattened upper end. Corroded and badly damaged in places, with a recent separation of part of the surface layer. Iron. Dimensions: 63.5 x 15 x 10 mm (Pl. I, 5, Fig. 5).

6. Part of a massive rod with both ends broken off. A very badly preserved piece with extreme corrosion and deep cracks on the surface. Iron. Length: 111 mm. It was found together with no. 7, no. 11, and no. 26 (Pl. I, 6, Fig. 6).

7. Part of a flattened and corroded bar, broken on both ends, with the beginnings of a broadened section (?) on the lower edge. Iron. Length: 89 mm. Width: 21 and 25 mm. (Pl. I, 7, Fig. 7).

8. A spike with an irregular thickened upper part, a thick shank, and a pointed lower ending. Highly corroded, the surface damaged with cracks. Length: 130 mm (Pl. I, 8, Fig. 8).

9. A bent spike (?) with an irregularly thickened, almost oval upper section and a blunt lower end. Badly damaged, with layers of corrosion. Iron. Diameter of the upper section: 23 and 28 mm. Length: 108 mm. It was found with no. 24 and no. 64 (Pl. I, 9, Fig. 9).

10. Arrowhead, badly damaged by deep corrosion, broken off on the lower end. Iron. Length: 48 mm (Pl. I, 10, Fig. 10).

11. A nail with an irregular gently rounded oval head. Highly corroded, the surface layer broken off in several places. Iron. Diameter of the head: 34 and 36 mm. Length: oko 180 mm (Pl. II, 11, Fig. 11).

12. A nail with an oval widely hammered head and a relatively well preserved surface with a patch of corrosion in the middle of the shank. Iron. Diameter of the head: 35 and 30 mm. Length: 114 mm. It was found with no. 18 (Pl. II, 12, Fig. 12).

13. A nail with an irregular head with a rounded surface, rusted to the extent of separation and detachment of pieces of the surface layer. Iron. Diameter of the head: 23 x 19 mm. Length: 106 mm. It was found with no. 27 (Pl. II, 13, Fig. 13).

14. A nail with an irregular thickened circular head and a long shank with a broken lower end. The surface damaged by corrosion is covered in places by clumps of rust. Iron. Length: 93 mm (Pl. II, 14, Fig. 14).

15. A nail with an irregular rounded head, curved shank, and corroded surface. Iron. Length: 78 mm. It was found with no. 34 and no. 43 (Pl. II, 15, Fig. 15).

16. A nail with an asymmetric circular head, broken end of the shank, covered with a layer of corrosion. Iron. Length: 54 mm (Pl. II, 16, Fig. 16).

17. A nail with an oval rounded head, corroded surface, and a well-preserved metal core. Iron. Length: 87 mm. It was found with no. 19 (Pl. II, 17, Fig. 17).

18. A nail with an irregular, perhaps originally square head, the upper part broken off with a surface below a corroded layer. Iron. Length: 63 mm (Pl. II, 18, Fig. 18).

19. A nail with a large flat hammered irregular circular head, broken off in the lower part of the shaft, the surface badly corroded. Iron. Length: 47 mm (Pl. II, 19, Fig. 19).

20. A nail with a large flat square head, the lower part of the shank broken off, the surface covered with a rusted layer. Iron. Length: 50 mm (Pl. II, 20, Fig. 20).

21. A nail with a hammered oval head with irregular edges, under a thick corroded layer. Iron. Length: 41 mm (Pl. II, 21, Fig. 21).

22. The shank of a nail with traces of a break in the upper end; the surface covered by a thin layer of rust. Iron. Length: 47.5 mm. It was found with no. 22, no. 92, and no. 93 (Pl. II, 22, Fig. 22).

23. A piece of burnt iron, almost missing its core, with tiny pieces of charcoal remaining on an uneven surface covered by rust. Iron. Dimensions: 45 x 35 mm (Pl. II, 23, Fig. 23).

24. A fragment of iron slag with a porous cratered surface. Dimensions: 28 x 22 mm (Pl. II, 24, Fig. 24).

25. A melted porous clump of iron slag and some other grey coloured metal (?); pieces of a red-brown shiny coating were preserved in some hollows in a surface covered with rust, while other hollows contained burnt wood. Iron and some other melted and hardened metal (?). Dimensions: 93 x 68 x 50 mm (Pl. II, 25, Fig. 25).

26. Part of an arched bar with a marked lengthwise ridge that merges into an irregular circular protrusion on the top of the widening on the right end. A lengthwise edge is also emphasized along the outer edge on the reverse of the bar. A white metal under grey and dark patina, most probably lead. Length: 35 mm (Pl. III, 26).

27. A bent bar of unequal width broken in the middle, with damaged edges and an emphasized central ridge. Broken off on both ends. A white metal under a greyish coating, most probably lead. Length: ca. 52 mm (Pl. III, 27, Fig. 27).

28. A piece of amorphous white metal, rounded on one side and flat on the other, covered with a dirty dark and light grey coating. Lead. Dimensions: 25 x 15 mm (Pl. III, 28, Fig. 28).

29. A shapeless piece of melted, well preserved, shiny white metal, covered by a dirty grey dusty coating. Lead. Dimensions: 60 x 42 mm (Pl. III, 29, Fig. 29).

30. An irregular piece of sheet metal with the reverse covered by a dark grey coating; the front side is a paler grey, mostly mottled with white patches. Hard white metal. Silver (?). Dimensions: 39 x 31 mm (Pl. III, 30, Fig. 30).

31. A ring with a wide smooth closed hoop, its surface on both sides almost entirely covered by patches of green patina; A shiny while metal is visible in several places on the outer side. Silver (?). Diameter: 23 mm. It was found with no. 52 (Pl. III, 31, Fig. 31).

32. A simple ring with a closed hoop, completely covered with a layer of rough green patina. Bronze. Diameter: 21 mm. It was found with the following no. 33 (Pl. III, 32, Fig. 32).

33. A ring made from a piece of smooth wire whose ends were soldered to a central tiny sheet metal loop; the surface is shiny, darkened to brown, and mostly covered by a dusty green patina. Bronze. Diameter: 19 and 20 mm (Pl. III, 33, Fig. 33).

34. A ring with overlapping and joined ends made from a wire shaped on the outside with some instrument into alternating smooth rounded and densely ribbed sections, covered entirely with a green patina. Bronze. Diameter: 18 and 18.5 mm (Pl. III, 34, Fig. 34).

35. A ring with a banded hoop that forks in front into two arms on both sides, onto which was soldered a rosette shaped platelet composed of six circular elements with recesses for inserts: the outer recesses contained glass or stone elements of a pale blue colour, while the inserted element in the seventh central recess was dark red. The ring has green patina all over, the hoop is deformed, both forked arms are broken off on the right side, and one on the left. Diameter: 29 and 22 mm. (Pl. III, 35, Fig. 35).

36. An earring with an open circlet with distant ends and a loop formed by bending the wire; entirely covered by a green dusty patina. Bronze. Diameter: 27 and 22 mm (Pl. III, 36, Fig. 36).

37. An incomplete fibula with a banded bow and a reinforced lengthwise central rib, which originally bore a relief decoration with a currently unidentifiable motif. The foot was divided by transverse moulding, while the flat head was smooth and without decoration. The uneven surface of the clasp is covered by a dirty grey-brown and green patina, the edges are damaged, the bent bow is broken in the centre, and all parts of the fastening mechanism are missing: the lower part of the foot with the catch-plate, the pin, and the tubular shaft below the head with a wire axle for mounting the pin. Bronze. Length: 38 mm. Height: 24 mm. It was found with no. 58 and no. 59 (Pl. III, 37, Fig. 37).

38. A pendant with an elongated rhomboid plate, a drop-shaped lower element, and a loop for suspension made of wire drawn from the remainder of metal on the upper end with the rest coiled in six spirals below the loop. The platelet has a dotted linear decoration. The pendant is covered all over with a compact green patina; the loop is broken. Bronze. Length: 50 mm (Pl. III, 38, Fig. 38).

39. A piece of damaged, partly distorted sheet metal, broken on the upper end, with lateral indentations and an extended lower section with a rounded edge; remains of serration can be seen along the edge. The surface is covered with a compact green patina. Bronze. Dimensions: 31.5 x 26.5 mm (Pl. III, 39, Fig. 39).

40. A flat circular mount with two tangs for attachment on the reverse. The edge is divided by rounded elements, with a grooved furrow inside it where part of a fill of some black material was preserved in a short section on the left side. Below the green patina, which accumulated in layers on the reverse, the front side bore visible remains in a square border of a white metal coating, in places with a deep shine. The centre had shallow irregular depressions arranged in a rosette shape for some fill (?), which is difficult to differentiate with certainty from damage. A dilapidated piece, in the upper section pierced and torn, the edge damaged. Bronze. Diameter: 42 and 43 mm (Pl. III, 40, Fig. 40).

41. A fragment of a flat fitting with a perforated decoration, formed on the basis of a horizontal bar and two thin lower ones enclosing a triangular opening. The horizontal bar continues to a final triangular plate, while on the other end it is broken off, as are both of the lower bars, from which it continues at an angle. A thick accumulation is located on the rear side of the

triangular plate, which might indicate the remains of a pin or small rivet for attachment. The fragment is completely covered by an uneven thick layer of green patina. It was burnt. Bronze. Dimensions: 33 x 12 mm. It was found with the following no. 42. (Pl. III, 41, Fig. 41)

42. A piece of a flat fitting with perforated decoration, the remaining part formed into a pair of drop-shaped elements on a curved bar, with remnants of a continuation. It is covered by a thick layer of green patina with a pitted, grainy surface. It was exposed to fire. Bronze. Dimensions: 25 x 11 mm. (Pl. III, 42, Fig. 42)

43. A cast flat piece of metal, broken off on the upper narrower end and widening below. The other edges are preserved in their original form. Both surfaces are smooth and shiny, and the only decoration is a perpendicular dotted spiral. It is covered equally on both sides with a dark grey patina with a blue shine and scattered irregularly circular spots of thick rough green patina, which in the lower parts covers large areas. Bronze. Length: 54.5 mm. (Pl. III, 43, Fig. 43)

44. A hollow globular button with a circular loop for suspension. In the interior, below the loop, a transverse platelet is visible. It is badly damaged, with a large break in the wall. The surface is covered with a thick dirty grey greenish rough layer, while the loop for suspension has a green patina. The interior surface is whitish and shiny. The object was exposed to fire. Bronze (?). Height: 16 mm. Diameter: 14 mm. It was found with no. 195. (Pl. III, 44, Fig. 44)

45. Six identical hollow cast globular buttons with a pair of regular circular openings, located on both sides below the soldered oval wire loop. They were well made and exceptionally well preserved, only one being slightly damaged. They have their original smooth surface, covered by a thin compact patina with a blue-green nuance. Bronze. Height: 22 - 23 mm. Diameter: ca. 19 mm. (Pl. III, 45, Fig. 45)

46. A pair of flat appliqués in the form of a trefoil with identical figural depictions executed in bas-relief on the front. The unevenly formed reinforced edge that follows the contours of the trefoil contains a dolphin with an eye marked and an emphasized snout in the form of a beak, while the tail fin is lost in the folds of the head covering of a male head in right profile. A second female (?) left-facing profile is visible on the appliqué in a reversed position, and was stamped so that the nose of the first profile represents the chin of the other, and vice versa. Above the second profile were rounded pleats of draperies. Opposite the profiles, on the right side, the space was filled – so it seems – with a stylized floral motif in the form of a heart-shaped leaf with a flower above it. The image was precisely and well stamped, but the details can only be poorly recognized in the present worn state. The second example is worse preserved, while the depiction can be recognized on the basis of the first one. The appliqués have a grey-green patina on both

sides: on the front side the patina was retained in depressions, while the protruding parts of the relief were shiny and darkened to brown. Shiny fragments of a white metal are visible in places on both sides – the remains of silver plating, tinning, or the coating of the surface with some white alloy. The spot can clearly be seen on the rear of the appliqué where the element for attachment was broken off – a nail or rivet or traces of soldering. Bronze of a reddish colour or copper. Dimensions: 19 x 17 mm. (Pl. III, 46, Fig. 46)

47. A solid cast handle with a braided band, which forks in the upper section into two arms with pointed ends. The lower end of the handed handle depicts a male bearded and moustached mask with emphasized ears, large eyes, and heavy eyebrows. On the back of the mask and on the reverse of the arms, remains are preserved of a white metal with which the handle was soldered onto the neck of a vessel below the rim. The well-preserved piece has a greyish matte surface with scattered thick spots of green patina. Bronze. Height: 62 mm. Width: 23 mm. (Pl. IV, 47, Fig. 47)

48. A cast hook hung from a hoop with joined ends, which retained a piece of a chain braided from wire circlets. The hook and hoop are in good condition, covered with a brown-grey-green patina, while the remains of the chain are corroded and under a thick layer of patina. Bronze. Length of the hook: 42 mm. Length of the chain remnants: 13 mm. Diameter of the hoop: 21 mm. (Pl. IV, 48, Fig. 48)

49. A clamp with a circular loop for hanging and with long arms for driving in and bending. The original end on one arm is bent outwards, and is broken off on the other. The surface is uniformly covered with a compact green patina. Bronze. Length: 70 mm. (Pl. IV, 49, Fig. 49)

50. A piece of slightly bent hammered sheet metal with the original side edges and a distinct central lengthwise ridge, broken off on the upper and lower ends. It is covered by a grey-green patina with reddish spots. Bronze. Dimensions: 42.5 x 17.5 mm. (Pl. IV, 50, Fig. 50)

51. A piece of a worn and bent sheet metal band with a green patina. Bronze. Length: 43 mm. It was found together with no. 62 and no. 63, and the bronze remains on Fig. 55a, with the exception of the small piece on the lower left. (Pl. IV, 51, Fig. 51)

52. A piece of thin deformed sheet metal with a brown and green patina. Bronze. Dimensions: 46 x 24 mm. (Pl. IV, 52, Fig. 52)

53. A deformed platelet of thick sheet metal with damaged edges, broken into two parts, with a green patina. It was exposed to fire. Bronze. Length: 44 mm. (Pl. IV, 53, Fig. 53)

54. A platelet with patches of melted and solidified metal on the back, and with crumbs of charcoal left in the green patina layer. It had burnt. Bronze. Dimensions: 28 x 20 mm. It was found together with no. 71, no. 73, and one clump on Fig. 55a – the first on the lower left. (Pl. IV, 54, Fig. 54)

55. A piece of a thick rod (?) with one original preserved edge along the longer axis, cut off straight on the upper and lower ends. The surface, covered by a green patina, contains uneven clumps of metal under a coating the colour of iron rust. It had burnt. Bronze. Dimensions: 20 x 18 x 9 mm. Over ten bronze fragments (pieces of rods and bands, clumps of melted metal) of various dimensions (lengths from ca. 10 to 63 mm) share a similar extremely bad state of preservation, and in some cases only the green patina has survived without a trace of the metal core. (Pl. IV, 55, Fig. 55, Fig. 55a)

56. A needle with a rectangular eye in the upper flatly hammered section with damaged edges. The lower end of the shank is broken off. The well-preserved surface is coated with a thin blue-green patina. Bronze. Length: 136 mm. (Pl. IV, 56, Fig. 56)

57. The shank of a needle, judging from the traces of the lower section of the eye below the point of breakage in the thin upper section. Broken and bent, covered with a grey-green patina. Bronze. Length: 64 mm. (Pl. IV, 57, Fig. 57)

58. Part of a pin or needle with a green patina. Bronze. Length: 35 mm. (Pl. IV, 58, Fig. 58)

59. A bent section of a pin/needle, broken on both ends, with a green patina. Bronze. Length: 32 mm. (Pl. IV, 59, Fig. 59)

60. A bent section of a pin/needle, broken on both ends, with an uneven green patina. Bronze. Length: 45 mm. It was found together with the following no. 61. (Pl. IV, 60, Fig. 60)

61. The lower final section of a pin/needle, bent and broken, with a green patina. Bronze. Length: 26.5 mm. (Pl. IV, 61, Fig. 61)

62. A piece of badly corroded thick wire under a layer of green patina, in clumps in the upper section. Bronze. Length: 63 mm. (Pl. IV, 62, Fig. 62)

63. A fragment of a bar of uneven width, covered with a green patina. Bronze. Length: 28 mm. (Pl. IV, 63)

64. Part of a flat bar with an uneven surface covered with a green patina in which tiny pieces of charcoal remained. Bronze. Length: 34 mm. (Pl. IV, 64, Fig. 64)

65. A probe with a damaged scoop, covered with a dirty grey smooth coating, below which dusty green islands appear in places. Bronze. Length: 103 mm. (Pl. IV, 65, Fig. 65)

66. A probe with a damaged back, broken in the narrow section between the cylindrical handle and the scoop. Burnished bone. Length: 123 mm. (Pl. V, 66, Fig. 66)

67. A handle of a probe or the shank of a hairpin (?). Ground and polished bone. Length: 93 mm. (Pl. V, 67, Fig. 67)

68. A small tube with thin walls, broken off on both ends. The visible part of the inner surface is as smooth as the outer surface. Bone. Length: 49 mm. (Pl. V, 68, Fig. 68)

69. An animal bone, broken off on the upper end. Length: 41 mm. It was found with no. 177. (Pl. V, 69)

70. A fragment of a platelet with two faces, the rear smooth and glazed shiny, mottled with fine cracks like veins in marble, while the front has a matte surface. Two slightly lowered fields can be found on the front side: the upper is smooth, and the lower striped with dense ribs. The workmanship of the edging frame and the ribbing is very regular. A material ivory in colour with the hardness of stone. Stone? Recent porcelain? Dimensions: 28 x 17.5 mm. (Pl. V, 70, Fig. 70)

71. A relatively regular cube, which appears naturally formed or perhaps slightly worked. Stone with an anthracite hue. Dimensions: 16 x 16 x 12 mm. (Pl. V, 71, Fig. 71)

72. A flat fragment with the shape of a knife blade, with damaged lengthwise edges and upper layer chipped of in places, rounded on one end and broken off on the other. The surface of both sides is abraded in various directions; the scratches on the lower part of the surface are slanted. Stone with a grainy texture and a dark brown-purple hue. Length: 138 mm. (Pl. V, 72, Fig. 72)

73. A small fragment of painted wall plaster with preserved remains of coloured transparent bands: the upper pinkish, the lower greyish. Length: 28 mm. (Pl. V. 73, Fig. 73)

74. A rim fragment of a pottery vessel with a mostly preserved slip. On the inner side the slip is compact, somewhat shiny, with a uniform reddish hue, and outside it is matte brown and remained only in deeper sections, while it had peeled off protruding parts. Highly refined clay of considerable hardness. Dimensions: 16.5 x 15.5 mm. It was found with no. 111. (Pl. V, 74, Fig. 74)

75. A fragment of a clay lamp, broken at the sector at the transition from the disk to the thin walls, divided by annular mouldings. The remains of a relief decoration in the form of small arcades are preserved on the raised thickened edge of the disc. The surface of the fragment is covered by a high quality matter brown-orange slip. Refined soft clay with a powdery structure, pale ochre, no trace of admixture. Dimensions: 37 x 22 mm. (Pl. V, 75, Fig. 75)

76. A rim fragment of a simple vessel with coarse walls with a rough and blotchy grey-black surface. The clay is quite hard, without visible admixtures or impurities. It was exposed to fire. Dimensions: 34 x 25 mm. It was found with no. 154 and no. 173. (Pl. V, 76, Fig. 76)

77. A balsamarium with a rounded base with a small depression on its exterior side. The transition to the neck is marked by a slight indentation of the narrow receptacle, whose length is almost identical to the length of the neck. The edge of the rim is missing. Green glass. Dimensions: 68 x 19 mm. It was found together with the following no. 78. (Pl. VI, 77, Fig. 77)

78. A balsamarium like the preceding one, with a somewhat wider receptacle missing the upper section.

Green glass. Dimensions: 43 x 21 mm. (Pl. VI, 78, Fig. 78)

79. A cylindrical, slightly rounded neck of a balsamarium with a thickened everted rim and an almost horizontal beginning to the wall of the receptacle. Green glass. Dimensions: 30 x 27 mm. (Pl. VI, 79, Fig. 79)

80. A cylindrical neck of a balsamarium with a thickened rim slanted outwards and the beginning section of the walls of a rounded receptacle. Green glass. Dimensions: 39 x 31.5 mm. (Pl. VI, 80, Fig. 80)

81. The neck of a balsamarium with a thickened rim slanted outwards and the beginning section of the walls of a rounded receptacle. Green glass. Dimensions: 27 x 13 mm. It was found together with no. 105, no. 130, no. 189, and no. 202, along with another ten glass fragments of various colours and hues, with lengths from 20 to 46 mm. (Pl. VI, 81, Fig. 81, Fig. 81 a)

82. The neck of a balsamarium with a thickened, highly everted and downward slanted rim and the beginning section of rounded walls. Green glass. Dimensions: 38 x 34 mm. It was found together with no. 125 and no. 198, along with another green glass fragment and two colourless ones measuring from 28 to 32 mm. (Pl. VI, 82, Fig. 82, Fig. 82 a)

83. A partially preserved section of the thickened straight rim of a balsamarium with the beginning of the neck. Green glass. Dimensions: 37 x 13 mm. Diameter: 36 mm. It was found together with the next two no. 84 and no. 85, along with another 23 fragments of glass, measuring from 14 to 48 mm, all green, except for one colourless fragment and no. 213. h. (Pl. VI, 83, Fig. 83, Fig. 83 a)

84. Part of the neck of a balsamarium with a thickened and outward turned rim, whose edge is folded inwards. Green glass. Dimensions: 33 x 20.5 mm. (Pl. VI, 84, Fig. 84)

85. Part of a wide, thickened and outwardly turned rim, whose edge is attached to the inner side. A bit of the neck is preserved below the rim. Green glass. Dimensions: 50 x 23 mm. (Pl. VI. 85, Fig. 85)

86. Part of a wide rim, everted and attached to the inner side, with the beginning of the neck. Green glass. Height: 18 mm. Diameter: 41 mm. It was found together with no. 194. (Pl. VI, 86, Fig. 86)

87. Part of a neck with a slight indentation in the middle and the rim curved to the inner side. Green glass. Dimensions: 45 x 36 mm. It was found together with another six glass fragments of various colours, with lengths from 15 to 30.5 mm and no. 213i. (Pl. VI, 87, Fig. 87, Fig. 87 a)

88. A thickened rim curved to the inner side with the beginning of the walls of the vessel. The rim has a variable appearance and thickness, the layer of glass below the rim is doubly thickened in some placed, and the glass mass is full of air bubbles and stains. Green glass. Height: 22 mm. Width: 63 mm. Diameter of the rim: 56 and 53 mm. It was found together with no. 162. (Pl. VI, 88, Fig. 88)

89. A fragment of the neck section of a vessel with the rim curved to the inner side. Pale green glass. Height: 25 mm. Width: 22 mm. (Pl. VI, 89)

90. The upper part of a flask with a funnel-shaped rim and a cylindrical, slightly widened neck below. The decoration that begins below the slightly marked smooth moulding on the shoulders and probably covered the receptacle completely consists of rows of shallow slanted relief ribs. Of the thirty some remaining fragments from this vessel, only 4 or 5 have a smooth surface, the rest are ribbed. Amber glass. Height: 95 mm. Width: 100 mm. It was found together with a pile of tiny fragments of green glass, the largest with a length of 38 mm, as well as no. 213o. (Pl. VI, 90, Fig. 90, Fig. 90 a)

91. The neck of a jug with a slightly thickened trefoil rim. The preserved part of the banded handle has an unusual position. Green glass. Dimensions: 64 x 50 mm. (Pl. VII, 91, Fig. 91)

92. Part of the wall of a beaker with the flat part of the base on a shallow ringed foot. Above the base is a thicker part of the wall, smooth in the centre, of a different, somewhat thinner moulding, which divides it into two zones. In the upper zone, marked by an engraved line, the lower part was preserved of a decoration composed of a row of concave ground segments with an oval lower end. Colourless glass. Dimensions: 39 x 34 mm. Diameter of the base: ca. 45 mm. (Pl. VII, 92, Fig. 92)

93. A fragment of a beaker with a preserved decoration consisting of perpendicular, elongated, slightly curved ground ovals located beneath a horizontal engraved line. Colourless glass, appearing identical to that of the previous piece, with which it was found. Dimensions: 45.5 x 18.5 mm. (Pl. VII, 93, Fig. 93)

94. Part of the circular stem of a glass with a tubular ringed foot attached to the lower side of the base and with the lower part of the hollow stem. An extraneous long piece of glass remained on the reverse. Green glass. Diameter: 45 mm. It was found together with a fragment of milk white glass 28 mm long. (Pl. VII, 94, Fig. 94, Fig. 94a)

95. The rim section of a small bowl with one partially preserved vertical rib. Green glass. Dimensions: 48 x 37.5 mm. Diameter of the rim: ca. or more than 110 mm. (Pl. VII, 95, Fig. 95)

96. A rim fragment of a small bowl with the upper section of one thin vertical rib. Blue glass with a cobalt hue. Dimensions: 25 x 35 mm. It was found together with ten crystal-like small pieces of milky yellow glass with an alabaster appearance, the largest with a length of 17 mm. (Pl. VII, 96, Fig. 96, Fig. 96 a)

97. A fragment of a small bowl with part of the preserved rim and part of a vertical rib that is indented in the upper section. Green glass. It was found together with another two green and one colourless glass fragments, below and above 20 mm in length, and no. 213b. (Pl. VII, 97, Fig. 97, Fig. 97 a)

98. A fragment of the walls of a small bowl with a preserved central section of a vertical rib. Green glass.

Dimensions: 40 x 26 mm. It was found together with no. 155. (Pl. VII, 98, Fig. 98)

99. A fragment of a small bowl with the beginning part of a vertical rib. Green glass. Dimensions: 25 x 21.5 mm. (Pl. VII, 99, Fig. 99)

100. A fragment of a small bowl with part of a vertical rib. Green glass. Dimensions: 27 x 17 mm. It was found together with no. 143, along with five fragments of green glass and a broken fragment with an olive hue, from 20 to 52 mm in length. (Pl. VII, 100, Fig. 100, Fig. 100 a)

101. A fragment of a small bowl with the beginning part of a vertical rib. Green glass. Dimensions: 27 x 16, 5 mm. (Pl. VII, 101, Fig. 101)

102. A fragment of a small bowl with part of a vertical rib. Green glass. Dimensions: 31 x 26 mm. It was found together with no. 99, no. 103, no. 107 - no. 109, no. 119, no. 124, no. 137, and no. 185, along with another fourteen fragments of green glass, from 14 to 54 mm long, as well as no. 213l and no. 213n. (Pl. VII, 102, Fig. 102, Fig. 102 a)

103. A fragment of a small bowl with part of a vertical rib. Green glass. Dimensions: 25 x 21.5 mm. (Pl. VII, 103, Fig. 103)

104. One rim fragment and two wall fragments of a plate, the larger fragment broken into three pieces. The rim is rounded and turned outwards. The surface is flat on both sides, rough to the touch, with no shine and partly opalescent. Polychrome mosaic glass with a combination of yellow, green, and ivory. On the recent breaks of the largest fragment, along with more intensive green and yellow layers, layers of a bright red colour are visible in a few places. Dimensions of the rim section: 41.5 x 27.5 mm; larger piece: 66 x 41 mm; smaller piece: 24.5 x 15.5 mm. Diameter of the rim: around or more than 20 mm. (Pl. VIII, 104, Fig. 104)

105. A rim fragment of a vessel with part of an annular moulding on the lower side. Pale green glass. Dimensions: 43 x 16.5 mm. Diameter of the rim: around or more than 130 mm. (Pl. VIII, 105, Fig. 105)

106. Part of a vessel with a tubular outward turned rim bent onto the lower side. Amber glass. Dimensions: 63 x 14.5 mm. Diameter of the rim: ca. or more than 100 mm. (Pl. VIII, 106, Fig. 106)

107. Part of a tubular rim bent onto the outer side with the beginning of the walls. Blue glass with cobalt hue. Dimensions: 45 x 11 mm. Diameter of the rim: ca. or more than 100 mm. It was found together with no. 108 and no. 109. (Pl. VIII, 107, Fig. 107)

108. A small fragment of a tubular rim of decayed glass of an unclear colour. At the break point it is covered with a green opalescent coating while almost all the remaining surface is covered with a layer of pale orange colour without shine. Effects of burning? Dimensions: 29.5 x 7 mm. (Pl. VIII, 108, Fig. 108)

109. A small piece of a tubular rim with a small section of the wall. Green glass. The outer surface is covered with an opalescent membrane with a silvery gleam, while the inner cavity is lined with a pale yellow layer. Burnt? Dimensions: 16 x 10 mm. (Pl. VIII, 109, Fig. 109)

110. A fragment of a vessel with a tubular rim, curved towards the interior side, with the beginning part of the wall. Green glass. Dimensions: 40.5 x 15.5 mm. It was found together with no. 160, along with a piece of green glass 27 mm long. (Pl. VIII, 110, Fig. 110, Fig. 110 a)

111. Part of a vessel with a thickened rim, turned outwards and curved and attached inside, with the beginning of the wall. Green glass. Dimensions: 40.5 x 19.5 mm. It was found together with the pottery fragment no. 74, as well as no. 213k. (Pl. VIII, 111, Fig. 111)

112. Part of a vessel with a thickened, inwards bent rim and the beginning of the wall. Blue-green glass. Dimensions: 31 x 15 mm. It was found together with no. 126, no. 186, and no. 187, along with another two pieces of glass with a similar hue, 16 and 24 mm long. (Pl. VIII, 112, Fig. 112, Fig. 112 a)

113. A fragment of a vessel with a thickened rim, slightly curved towards the inside, and the straight beginning of part of the wall. Green glass. Dimensions: 34.5 x 17 mm. It was found together with no. 128 and no. 140, and another two pieces of glass: blue and green grooved and curved (probably from the neck of a balsamarium), 48 and 29 mm long. (Pl. VIII, 113, Fig. 113, Fig. 113 a)

114. A fragment of a massive vessel with thick walls and a broad everted and flattened rim. Green glass. Dimensions: 33.5 x 20 mm. (Pl. VIII, 114, Fig. 114)

115. A fragment of an incomplete, broad and slanted rim with part of the wall. The inside edge of the rim is broken off along the entire length. Green glass. Dimensions: 42 x 27 mm. It was found together with another broken, thick, approximately semicircular, spoon-like indented piece of greenish glass of crystalline structure, dimensions 20 x 13 mm. (Pl. VIII, 115, Fig. 115, Fig. 115 a)

116. A partly preserved, broad and flattened rim with the beginning of the wall. Green glass. Dimensions: 37.5 x 10 mm. Diameter of the rim: 56 mm. It was found together with no. 120, along with another four pieces of green glass and one piece of colourless glass from 27 to 59 mm long, and also an animal bone 55 mm long, and the fragments no. 213nj, no. 213lj, and no. 213m. (Pl. VIII, 116, Fig. 116, Fig. 116 a)

117. A fragment of a vessel with smooth walls and a rim thickened inside. Colourless glass with the appearance of alabaster. Dimensions 36 x 25 mm. It was found together with no. 138, and also two pieces of green glass and two pieces of colourless glass from 11 to 29 mm long. (Pl. VIII, 117, Fig. 117, Fig. 117 a)

118. Part of a vessel with a slightly emphasized and thickened rim. Green glass. Dimensions: 40 x 27 mm. It was found together with no. 149 and no. 157, as well as: a shattered fragment of the wall of a vessel of dark blue glass measuring 48 x 28 mm, a piece of amber glass 30 mm long, and three small pieces of

green glass 14, 16, and 17 mm. (Pl. VIII, 118, Fig. 118, Fig. 118 a)

119. Part of a vessel with a simple rounded rim and horizontal thickening on the preserved part of the wall. Green glass. Dimensions: 29 x 23 mm. (Pl. VIII, 119, Fig. 119)

120. A fragment of a vessel with curved walls and a thickened rim slanted towards the outside. Green glass. Dimensions: 22.5 x 13 mm. (Pl. VIII, 120, Fig. 120)

121. A fragment with a thickened rim and straight walls. Pale green glass. Dimensions: 13.5 x 15 mm. It was found together with no. 123, no. 153, no. 164, no. 178, no. 180, no. 197, and no. 213, as well as thirteen glass fragments of various colours, from 10 to 60 mm long. (Pl. VIII, 121, Fig. 121, Fig. 121 a)

122. A fragment of a smooth slightly curved wall with a rim thickened on the inside. Green glass. Dimensions: 37 x 14 mm. It was found together with no. 146 and no. 192, and also another nine pieces of green and colourless glass from 10 to 38 mm long. (Pl. VIII, 122, Fig. 122, Fig. 122 a)

123. A fragment of a straight rounded rim. Green glass. Length: 32 mm. (Pl. VIII, 123, Fig. 23)

124. A small fragment of a vessel with a slightly emphasized and thickened rim. Pale blue glass. Dimensions: 15 x 7 mm. (Pl. IX, 124, Fig. 124)

125. A fragment of the thickened straight rim and smooth wall. Green glass. Dimensions: 19 x 15.5 mm. (Pl. IX, 125, Fig. 125)

126. A fragment of a vessel with a slightly thickened, rounded, out-turned rim. Blue-green glass. Dimensions: 27 x 11 mm. (Pl. IX, 126, Fig. 126)

127. A fragment of a vessel with a rounded rim and horizontal moulding on the exterior of the wall. Intensively green glass. Dimensions: 21.5 x 15 mm. It was found together with another piece of green glass 18.5 mm long, two broken corner pieces of pale blue glass 10 and 11 mm long, and no. 213e. (Pl. IX, 127, Fig. 127, Fig. 127a)

128. A fragment of a vessel with a rim sloping outwards and broken off straight. The exterior below the rim is decorated with horizontal shallow blue-white glass threads, which cannot be seen in profile. Dark blue cobalt glass. Dimensions: 19.5 x 19 mm. (Pl. IX, 128, Fig. 128)

129. A piece of a vessel with a straight rim sloping inwards and a smooth curved wall. Green glass. Dimensions: 32 x 28 mm. It was found together with another eight pieces of colourless, amber, and green glass from 17.5 to 50 mm long. (Pl. IX, 129, Fig. 129, Fig. 129 a)

130. Part of a vessel with a flat rim and smooth curved walls. Olive green glass. Dimensions: 22.5 x 22.5 mm. (Pl. IX, 130, Fig. 130)

131. Part of a vessel with a flat rim and smooth walls. Olive green glass. Dimensions: 25 x 22 mm. It was found together with no. 169 and no. 170, as well as three fragments of colourless and green glass from 33 to 37 mm long. (Pl. IX, 131, Fig. 131, Fig. 131a)

132. Part of a vessel with smooth curved walls. Colourless glass. Dimensions: 25 x 22 mm. It was found together with no. 150, no. 166, and no. 190, along with fourteen fragments of colourless, amber and green glass from 20 to 55 mm long. (Pl. IX, 132, Fig. 132, Fig. 132a)

133. Part of a very thin wall with uneven shallowly impressed double concentric circles on the outer surface, which are slightly convex on the inner side. Colourless glass. Dimensions: 19.5 x 14 mm. It was found together with br. 152. (Pl. IX, 133, Fig. 133)

134. Part of a vessel with thin walls and a curved white trail located beneath an irregular break on the upper end. Colourless glass. Dimensions: 23 x 18 mm. It was found together with no. 135, no. 145, no. 176, no. 191, and no. 214, along with another eight fragments of primarily colourless glass from 12 to 40 mm long, and a small piece of thin grey pottery with a grooved surface, 25 mm long. (Pl. IX, 134, Fig. 134, Fig. 134a)

135. The curved part of a foil thin wall of a vessel. Colourless milk glass. Dimensions: 27 x 17 mm. (Pl. IX, 135, Fig. 135)

136. Part of a vessel with smooth walls and a straight rim. Colourless glass. Dimensions: 54 x 44 mm. (Pl. IX, 136, Fig. 136)

137. Part of the wall of a vessel with a pair of impressed irregular white islands on the outer surface. Blue-grey milky opaque glass. Dimensions: 21 x 15 mm. (Pl. IX, 137, Fig. 137)

138. A fragment of the wall of a vessel with the beginning section of the base, broken into two parts. Green glass. Dimensions: 45 x 43 mm. (Pl. IX, 138, Fig. 138)

139. Part of a vessel with curved walls and an applied web of relief ribs on the outer surface. Broken into three parts. Green glass. Dimensions: 54.5 x 48 mm. It was found together with no. 212, as well as another fifteen fragments of primarily colourless flat glass, from 22 to 66 mm long. (Pl. IX, 139, Fig. 139, Fig. 139a)

140. Part of a vessel with a preserved small section of the base. On the outer side of the lowered rounded walls are relief ribs that narrow towards the top. Colourless greenish-hued glass. Dimensions: 68 x 52 mm. (Pl. IX, 140, Fig. 140)

141. A tubular piece, narrowed below and broken off on both ends. Greenish glass. Length: 39 mm. Diameter: 7 and 3 mm. It was found together with no. 175 and no. 181, as well as another thirteen fragments of colourless and green glass from 15 to 51.5 mm long. (Pl. IX, 141, Fig. 141, Fig. 141a)

142. An irregular bent piece of wall with indentations on the surface and with a fused triangular piece of glass on the front side. Green glass. Dimensions: 34 x 24 mm. It was found together with no. 207, as well as three pieces of colourless and green glass from 20 to 43 mm long and a brown fragment 43 mm long. (Pl. IX, 142, Fig. 142, Fig. 142a)

143. A fragment of the globular wall of a vessel with the beginning of the neck and with a thickened lower section, broken into two pieces. Green glass. Dimensions: 64 x 59 mm. (Pl. X, 143, Fig. 143)

144. A fragment of a vessel with rounded walls and part of a slightly indented base, broken into four pieces. The glass contains tiny and larger air bubbles. Green glass. Dimensions: 70 x 30 mm. It was found together with another pair of green pieces of glass, 17 and 24 mm long. (Pl. X, 144, Fig. 144, Fig. 144a)

145. A partially preserved thickened base of a vessel with a ringed foot. Green glass. Height: 15 mm. Diameter: 37.5 mm. (Pl. X, 145, Fig. 145)

146. Part of the thickened base of a vessel with a ringed foot. Green glass. Dimensions: 21.5 x 17 mm. Diameter: 60 mm. (Pl. X, 146, Fig. 146)

147. A piece of a base with a ringed foot. Green glass. Dimensions: 17.5 x 12 mm. It was found together with eight fragments of colourless, amber, and green glass, from 14 to 30 mm long. (Pl. X, 147, Fig. 147, Fig. 147a)

148. A partially preserved thickened base with a tubular ring-shaped fulcrum folded onto the upper side. Green glass. Dimensions: 43 x 6 mm. (Pl. X, 148, Fig. 148)

149. Part of a ringed foot with the beginning of the base and wall (?) preserved. Green glass. Dimensions: 27 x 14 mm. (Pl. X, 149, Fig. 149)

150. An irregular piece of a base with a ringed foot (?), broken off on all sides. Greenish glass. Dimensions: 30 x 21 mm. (Pl. X, 150, Fig. 150)

151. Part of a thickened base with a tubular ring bent upwards and the beginning of the walls. Colourless glass. Dimensions: 29 x 20 mm. Diameter: ca. 65 mm. It was found together with no. 196. (Pl. X, 151, Fig. 151)

152. Part of a base with a ringed foot and the beginning of a thinned wall. Green glass. Dimensions: 18 x 15 mm. Diameter: 35 mm. (Pl. X, 152, Fig. 152)

153. Part of a base with a ringed foot and part of the wall. Colourless greenish-hued glass. Dimensions: 51 x 22 mm. Diameter of the base: 45 mm. (Pl. X, 153, Fig. 153)

154. A partially preserved base with a ringed foot and part of the wall. Colourless greenish-hued glass. Dimensions: 35.5 x 27.5 mm. Diameter: 40 mm. It was found together with five fragments of colourless, amber, and green glass, from 25 to 49 mm long. (Pl. X, 154, Fig. 154, Fig. 154a)

155. Part of a base with a ringed foot. Amber glass. Dimensions: 40 x 26 mm. Diameter: 100 mm. (Pl. X, 155, Fig. 155)

156. An irregular circular base to a vessel, with a slightly depressed centre and a ringed foot. Green glass. Height: 7.5 mm. Diameter: 38 and 36.5 mm. It was found together with no. 184, as well as six fragments of colourless and greenish glass, from 10 to 39 mm long. (Pl. X, 156, Fig. 156, Fig. 156a)

157. A partially preserved, slightly hollowed base with a ringed foot and the beginning of the wall. Cobalt blue glass. Dimensions: 51 x 11 mm. Diameter of the base: 53 mm. (Pl. X, 157, Fig. 157)

158. Part of a slightly sagging and thickened base with a ringed foot, broken into two parts. Colourless glass. Dimensions: 31.5 x 15 mm. Diameter: 65 mm. It was found together with three fragments of colourless and green glass, 16, 33, and 44 mm long. (Pl. X, 158, Fig. 158, Fig. 158a)

159. A fragment of a ringed foot with a piece of the base. Green glass. Dimensions: 25 x 14.5 mm. It was found together with a pair of fragments of green glass 26 and 40 mm long. (Pl. X, 159, Fig. 159, Fig. 159a)

160. Part of a slightly sagging base with a ringed foot. Green glass. Dimensions: 34 x 17.5 mm. (Pl. XI, 160, Fig. 160)

161. Part of a base with a massive foot straight from the bottom or part of a wide level rim. Green glass. Dimensions: 40 x 21 mm. It was found together with no. 172 and no. 211. (Pl. XI, 161, Fig. 161)

162. The lower part of a vessel with a flat thickened base and perpendicular massive walls. Intensive green glass. Dimensions: 33 x 22.5 mm. (Pl. XI, 162, Fig. 162)

163. Part of a vessel with a flat thick base and slightly out-turned walls. Colourless greenish-hued glass. Dimensions: 41 x 24 mm. Diameter of the base: 65 mm. It was found together with no. 179, as well as three fragments of colourless glass 19, 20, and 61 mm long. (Pl. XI, 163, Fig. 163, Fig. 163a)

164. Part of a vessel with a thick flat base and perpendicular walls. Colourless greenish-hued glass. Dimensions: 38 x 33 mm. Diameter of the base: 85 mm. (Pl. XI, 164, Fig. 164)

165. Part of a vessel with a massive base and walls. Colourless greenish-hued glass. Dimensions: 49 x 25 x 17 mm. (Pl. XI, 165, Fig. 165)

166. Part of a vessel with a flat base and thin walls. Colourless glass. Dimensions: 25 x 21 x 14 mm. (Pl. XI, 166, Fig. 166)

167. Part of a vessel with a flat base and rounded walls. Colourless greenish-hued glass. Dimensions: 40 x 21.5 mm. Diameter: 85 mm. It was found together with a fragment of amber hue, 20.5 mm long. (Pl. XI, 167, Fig. 167, Fig. 167 a)

168. Part of a vessel with a flat base and thin walls curved on the outer side. Colourless glass. Dimensions: 41 x 19 mm. Diameter of the base: 40 mm. (Pl. XI, 168, Fig. 168)

169. Part of a vessel with a slightly indented base and rounded walls. Green glass. Dimensions: 33 x 19 x 9 mm. Diameter of the base: 50 mm. (Pl. XI, 169, Fig. 169)

170. Part of a vessel with a flat base and thin walls. Green glass. Dimensions: 64 x 41 x 24. Diameter of the base: 75 mm. (Pl. XI, 170, Fig. 170)

171. Fragment of a vessel with a thick indented base. The lower side of the base has two concentric relief ribs, the one nearer the centre interrupted. Green glass. Dimensions: 28 x 21 mm. It was found together with another three small fragments of green and colourless glass 31.5, 40, and 44 mm long. (Pl. XI, 171, Fig. 171, Fig. 171 a)

172. Fragment of a vessel with a square base with the

remains of one annular relief rib and the beginning of another on the outer side. Colourless greenish glass. Dimensions: 40 x 21 mm. It was found together with four small pieces of green and colourless glass, from 16 to 40 mm long. (Pl. XI, 172, Fig. 172, Fig. 172 a)

173. Fragment of a vessel with a thick, slightly indented base, with the remains of an annular relief ribs surrounding an incised square on the outer side. Dimensions: 63 x 41 mm. (Pl. XII, 173, Fig. 173)

174. Part of a vessel with a thickened base with a poorly centred relief square with indented sides on the outer side. The glass contained some tiny air bubbles. Green glass. Dimensions: 59 x 32 mm. It was found together with three fragments of green glass 20, 32, and 50 mm long. (Pl. XII, 174, Fig. 174, Fig. 174 a)

175. Part of a vessel with an irregular thickened base, broken into three pieces. Green glass. Dimensions: 66 x 55 mm. (Pl. XII, 175, Fig. 175)

176. Fragment of a vessel with a thick, slightly indented, probably square base, with the beginning of the thin walls on one side. Green glass. Dimensions: 62 x 40 mm. (Pl. XII, 176, Fig. 176)

177. Fragment of a vessel with a thick, probably square base. Green glass. Dimensions: 49 x 29 mm. It was found together with no. 69, and a small piece of colourless glass 14 mm long. (Pl. XII, 177, Fig. 177, Fig. 177 a)

178. Part of a vessel with a deeply indented base and straight, smooth walls thickened in the middle. Colourless glass. Dimensions: 58 x 35 mm. Diameter of the base: 35 mm. (Pl. XII, 178, Fig. 178)

179. A chipped fragment of a vessel with a deeply indented base. Green glass. Height: 13 mm. Diameter: 41 mm. (Pl. XII, 179, Fig. 179)

180. An irregular fragment of a vessel with an indented base. Colourless glass. Dimensions: 28.5 x 17.5 mm. (Pl. XII, 180, Fig. 180)

181. A smooth handle of semicircular section. It was fused to the vessel at a point below the thumb rest, which rose above the rim of the vessel. Opaque dark green and blue glass with red lengthwise threads. The glass remaining on the lower part of the handle is cobalt blue. Dimensions: 44 x 15.5 mm. (Pl. XIII, 181, Fig. 181)

182. A fragment of a smooth handle of semicircular section. The broadened lower section is concave like a spoon on the inner side. Dense vertical fissures are visible on the outer surface. Green glass of the same hue as the preserved remnant of the vessel wall. Dimensions: 32 x 26 mm. (Pl. XIII, 182, Fig. 182)

183. A fragment of a smooth banded handle with the preserved original left and lower edges, where a piece of the wall of the vessel was retained. Green glass. Dimensions: 20 x 14 mm. It was found together with another five fragments of green glass from 15 to 33 mm long and no. 213f. (Pl. XIII, 183, Fig. 183, Fig. 183 a)

184. A ribbon handle bent above the rim into a knee shape with reinforcing lengthwise ribs along the sides and parts of the wall of the vessel preserved on the ends. Green glass. Dimensions: 65 x 29 mm. (Pl. XIII, 184, Fig. 184)

185. A partially preserved ribbon handle with both ends broken but the original right lengthwise reinforcing rib preserved. Green glass. Dimensions: 50.5 x 21 mm. (Pl. XIII, 185, Fig. 185)

186. A fragment of the upper curved part of a ribbon handle with a section of the vessel. Green glass. Dimensions: 28 x 16 mm. (Pl. XIII, 186, Fig. 186)

187. The upper curved fragment of a ribbon handle reinforced with lengthwise edges and a small piece of the vessel. Green glass. Dimensions: 21 x 18 mm. (Pl. XIII, 187, Fig. 187)

188. A fragment of a broad banded handle, slightly folded vertically, with the left lengthwise and lower edges preserved, where part of the wall of the vessel was preserved. Green glass. Dimensions: 29 x 22 mm. It was found together with ten small pieces of colourless, green, and amber glass, from 18 to 51.5 mm long. (Pl. XIII, 188, Fig. 188, Fig. 188 a)

189. An incomplete broad short vertically ribbed handle with a uniform banded horizontal section. Broken off on the upper edge, chipped on the lower edge. The lengthwise edge sections are partly hollow and filled with impurities. The glass is stained in places, and excess drops of glass are visible on the surface. Olive green glass. Dimensions: 39 x 25 mm. (Pl. XIII, 189, Fig. 189)

190. A fragment of a banded handle (?), with a ribbed exterior surface. Olive green glass. Dimensions: 36 x 26 mm. (Pl. XIII, 190, Fig. 190)

191. A fragment of a ribbed ribbon handle (?), with the preserved section of one vertical edge. Colourless greenish glass. Dimensions: 42 x 24 mm. (Pl. XIII, 191, Fig. 191)

192. A tubular handle (?), bent like a knee in the narrower upper section, broken on both ends. Green glass. Dimensions: 42 x 16 mm. (Pl. XIII, 192, Fig. 192)

193. An arched, smooth handle with a thick banded section, broken on both ends. Casting seams run along the middle of the outer and inner sides. Dark green glass. Height: 42 mm. Width: 25 mm. Dimensions of the cross-section: 13 x 8 mm. It was found together with no. 209. (Pl. XIII, 193, Fig. 193)

194. Three irregular cubes with sharp edges. Green and blue glass. Length: 9 and 12 mm. (Pl. XIV, 194, Fig. 194)

195. Two irregular cubes with uneven edges. Cobalt blue glass. Length: 11 and 12 mm. (Pl. XIV, 195, Fig. 195)

196. Two irregular cubes. Green glass. Length: 10 mm. (Pl. XIV, 196, Fig. 196)

197. An irregular chipped cube. Blue glass. Length: 14 mm. (Pl. XIV, 197, Fig. 197)

198. A chipped cube. Green glass. Length: 11 mm. (Pl. XIV, 198, Fig. 198)

199. A half preserved thick circular disc. Milky white opaque glass. Diameter: 30 mm. (Pl. XIV, 199, Fig. 199)

200. A half preserved thick circular disc. Black opaque

glass. Diameter: 19 mm. It was found together with no. 201 and no. 204. (Pl. XIV, 200, Fig. 200)

201. A circular and highly rounded disc. White opaque glass. Diameter: 13.5 mm. (Pl. XIV, 201, Fig. 201)

202. A round bead with a tubular opening, broken into several small pieces. Green glass. Height: ca. 5 mm. Diameter: ca. 6 mm. (Pl. XIV, 202, Fig. 202)

203. A cylindrical bead with a tubular opening and a surface formed into six rectangular fields with gentle edges. Opalescent glass with emerald hues. Length: 9.5 mm. Diameter: 7 mm. It was found together with no. 205. (Pl. XIV, 203, Fig. 203)

204. A ribbed bead with slightly slanted narrow lobes. Green glass. Length: 11 mm. Diameter: 13 and 12 mm. (Pl. XIV, 204, Fig. 204)

205. An asymmetrical ribbed bead with irregularly arranged and carved grooves between the lobes. Blue-green glass under a dusty grey coating. Length: 13 mm. Diameter: 15 mm. (Pl. XIV, 205, Fig. 205)

206. Part of a smooth bracelet with uneven thickness and section, partially covered by opalescent stains in beige shades with golden sparkles. Black opaque glass. Length: 37 mm. Diameter of the hoop: around 70 mm. (Pl. XIV, 206, Fig. 206)

207. Part of a smooth bracelet, the outer surface dull and covered with tiny scratches, the inner surface partly opalescent. Black opaque glass. Length: 40 mm. Diameter of the hoop: around 50 mm. (Pl. XIV, 207, Fig. 207)

208. Part of a bracelet with carved decoration. The upper, outer, and lower surfaces are divided into there lengthwise bands: the edge bands are filled with a dense row of ribs, while the broad exterior band is divided into rectangular fields containing incised crossed grooves in the form of the letter X. Brown-black jet, rather dull. Length: 45.5 mm. Diameter of the hoop: around 60 mm. (Pl. XIV, 208, Fig. 208)

209. Two pieces of cobalt blue raw glass, partially opalescent. Dimensions: 40 x 33 mm and 26 x 23 mm. (Pl. XIV, 209, Fig. 209, Fig. 209a)

210. A piece of cobalt blue raw glass. Dimensions: 53 x 40 mm. (Pl. XIV, 210, Fig. 210)

211. A piece of opalescent cobalt blue raw glass. Dimensions: 31 x 30 mm. (Pl. XIV, 211, Fig. 211)

212. A piece of raw glass covered by an opalescent layer with emerald green and purple shades; a ruby red hue is visible on the thinner transparent edges. Dimensions: 33 x 21 mm. (Pl. XIV, 212, Fig. 212)

213. A piece of raw glass with one flat surface. Colourless refined glass with no admixtures. Dimensions: 55 x 31 mm. (Pl. XIV, 213, Fig. 213, Fig. 121a)

213a – 213d. Four pieces of raw glass, three found individually, and one with no. 97: three are cobalt blue, with the dimensions 21 x 17, 26 x 7, and 25 x 18 mm, while the fourth is milky green, measuring 37 x 18 mm. (Fig. 213a – 213d)

213e. Two broken pieces of pale blue glass 10 and 11 mm long and three small irregular squares of cobalt blue raw glass 10, 8, and 11 mm long. (Fig. 213e)

213f. A small piece of drop-shaped green glass 6 mm long. (Fig. 213f)

213g. A piece of stone, 41 x 20 x 25 mm, with one surface coated with a melted glaze-like shiny glass scoria of dark and light blue with a milky tone and tiny white dots – probably an admixture; the other surfaces are covered with a porous dirty grey burnt glass slag. (Fig. 213g)

213h. A piece of burnt glass slag 56 mm long. (Fig. 213h)

213i. A small piece of glass slag 22 mm long. (Fig. 213i)

213j. A clump of glass slag with a burnt core covered with scoria, 20 mm long. (Fig. 213j)

213k. A porous, 32 mm long, melted brown clump with a reddish layer of apparently iron rust and the hardness of stone, covered in places with glazed islands that when moistened sometimes take on the bluish grey shine of molten glass or slag. (Fig. 213k)

213l. A small piece of burnt glass slag 20 mm long. (Fig. 213l)

213lj - 213m. Two pieces of glass slag, 44 and 55 mm long, with small pieces of charcoal remaining in the hollows. (Fig. 213lj-213m)

213n. Three small pieces of charcoal 6 – 17 mm long. (Fig. 213n)

213nj. A small piece of carbonized wood with the original appearance preserved, 23.5 mm long and 16-17 mm in diameter. (Fig. 213nj)

213o. A small piece of charcoal and a small piece of dried (?) wood, 24 and 28 mm long. (Fig. 213o)

214. Twenty-one shapeless pieces of jet of various thickness and lengths from ca. 10 to 25 mm. The appearance of the wrinkled surface ranges from brown-black matte or lumpy to pure black and very shiny, while the section at breaks on all pieces is black and shiny. (Pl. XIV, 214, Fig. 214)

215. Four small pieces of jet with a black surface, 16, 19, 20, and 22 mm long. They were found together with two fragments of greenish glass 22.5 and 28 mm long. (Fig. 215, Fig. 215a)

216. Thirty variously coloured glass fragments found individually or in groups of two to eight pieces. (Fig. 216)

Abbreviations

ActaAHung	Acta Archaeologica Academiae Scientiarum Hungaricae (Budapest)
AH	Archaeologia Hungarica (Budapest)
AVes	Arheološki vestnik. Acta archaeologica. Slovenska akademija znanosti in umetnosti-Sekcija za arheologijo (Ljubljana)
BARIntSer	British Archaeological Reports, International Series (Oxford)
CCAVV	Corpus delle Collezioni Archeologice del Vetro nel Veneto. Giunta Regionale del Veneto. Comitato Nazionale Italiano-Association Internationale pour l'Histoire du Verre.
DissPann	Dissertationes Pannonicae (Budapest)
FolArch	Folia Archaeologica. A Magyar Nemzeti Muzeum (Budapest)
Instrumentum	Bulletin du Groupe de travail européen sur l'artisanat et les productions manufacturées dans l'Antiquité (Montagnac)
JahresberGPV	Jahresbericht. Gesselschaft pro Vindonissa (Brugg)
JRGZ	Jahrbuch des Römisch-Germanischen Zentralmuseums Mainz (Mainz)
KärntMussch	Kärntner Museumsschriften. Archäologische Forschungen zu den Grabungen auf dem Magdalensberg (Klagenfurt)
KatMonLj	Katalogi in monografije. Narodni muzej v Ljubljani (Ljubljana)
KatMonZ	Katalozi i monografije Arheološkog muzeja u Zagrebu (Zagreb)
Limesforsch	Limesforschungen. Römisch-Germanische Kommission des Deutschen Archäologischen Instituts (Berlin)
Opera	Opera Instituti Archaeologici Sloveniae. Znanstveno raziskovalni center Slovenske akademije znanosti in umetnosti-Inštitut za arheologijo (Ljubljana)
PWRE	Paulys Realencyclopaidie der Classichen Altertumswissenschaft (München)
Prilozi	Prilozi Instituta za arheologiju u Zagrebu / Contributions of Institute of Archaeology in Zagreb (Zagreb)
Situla	Situla. Razprave Narodnega muzeja v Ljubljani (Ljubljna)
VAMZ	Vjesnik Arheološkog muzeja u Zagrebu (Zagreb)
VeröffKommErf	Veröffentlichung der Kommission zur archäologischen Erforschung des spätrömischen Raetien der Bayerische Akademie der Wissenschaften (München)
VHAD	Vjesnik Hrvatskog arheološkog društva, nova serija (Zagreb)

BIBLIOGRAPHY

ANTIČKA BRONZA 1969:
Antička bronza u Jugoslaviji. Greek, Roman and Early Christian Bronzes in Yugoslavia – catalogue (Đ. Mano Zisi), National Museum Belgrade, Belgrade.

BAUER A. 1936:
Rimska olovna plastika s osobitim obzirom na materijal pohranjen u Hrvatskom narodnom muzeju. VHAD ns XVII, 1-35.

BIAGGIO SIMONA S. 1991:
I vetri romani provenienti dalle terre dell' attuale Canton Ticino. Vol. I, Locarno.

BISHOP M. C. 1988:
Cavalry equipment of the Roman army in the first century A. D. In Coulston (ed.), Military Equipment and the Identity of Roman Soldiers, BARintser 394.

BECATTI G. 1955:
Oreficerie antiche dalle minoiche alle barbariche. Roma.

BOLLA M. 1994:
Vasellame romano in bronzo nelle Civiche Raccolte Archeologiche di Milano. Rassegna di studi del Civico Museo Archeologico e del Civico gabinetto numismatico di Milano, Supplemento XI.

BOŽIČ D. 1999:
Tre insediamenti minori del gruppo protostorico di Idrija Pri Bači dell' Isontino. Università degli studi di Bologna Dipartimento di Archeologia, Studi e Scavi 8. Studio e Conservazione degli insediamenti minori romani in area Alpina. Atti dell' incontro di studi, Forgaria del Friuli, 20 settembre 1997 (S. Santoro Bianchi). University Press Bologna, 71-79.

BREŠČAK D. 1982:
Antično bronasto posodje Slovenije / Roman Bronze Vessels in Slovenia. Situla 22/1.

BRONZI ANTICHI 2000:
Bronzi antichi del Museo Archeologico di Padova. Statuette figurate egizie, etrusche, venetiche e italiche, armi preromane, romane e medioevali, gioielli e ogetti di ornamento, *instrumentum domesticum* dal deposito Museo – Catalogo della Mostra (G. Zampieri - B. Lavarone), Padova.

BRUNŠMID J. 1901:
Arheološke bilješke iz Dalmacije i Panonije.VHADns V, 87-168.

BURGER A. S. 1966:
The Late Roman Cemetery at Ságvár. ActaAHung XVIII.

BURKOWSKY Z. 1999:
Sisak u prapovijesti, antici i starohrvatskom dobu – catalogue. Gradski muzej Sisak, Sisak.

BURNUM 2007:
Grupa autora: Rimska vojska u Burnumu. L'esercito romano a Burnum. Katalozi i monografije Burnum 2. Drniš – Šibenik – Zadar.

CALVI M. C. 1969:
I vetri romani – Museo di Aquileia (Estratto da: I vetri romani del Museo di Aquileia 1968). Pubblicazioni dell' Associazione Nazionale per Aquileia 8, Aquileia.

CASAGRANDE C.- CESELIN F. 2003:
Vetri antichi delle Province di Belluno, Treviso e Vicenza. CCAVV 7.

CZURDA RUTH B. 1979:
Die Römischen Gläser vom Magdalensberg. KärntMussch 65.

DA AQUILEIA AL DANUBIO 2001:
Da Aquileia . . . al Danubio – Materiali per una mostra (M. Buora). Civici Musei di Udine, Comune di Udine, Museo Archeologico. Archeologia di frontiera 4, Trieste.

DEIMEL M. 1987:
Die Bronzekleinfunde vom Magdalensberg. KärntMussch 71.

DRENJE 1987:
Drenje – Rezultati istraživanja 1980. – 1985 / Research results 1980 – 1985 (Ž. Škoberne - R.Makjanić - R. Koščević). Publikacije Muzeja u Brdovcu sv. 1, Muzej Brdovec, Brdovec.

FABER A. 1973:
Građa za topografiju antičkog Siska. VAMZ 3, VI-VII (1972), 133-159.

FADIĆ I. 2002:
Antičke staklarske radionice u Liburniji. Godišnjak Centra za balkanološka ispitivanja Akademije nauka i umjetnosti B-H knj. 30, Sarajevo, 385-405.

GORENC M. 1979 - 1980:
Rimski olovni figuralni pečati iz Siska – exhibition leaflet. Arheološki muzej u Zagrebu, Zagreb.

GREGL Z. 1982:
Rimski medicinski instrumenti iz Hrvatske I / Die römischen medizinischen Instrumente aus Kroatien I. VAMZ 3. ser. – sv. XV, 175 – 198.

HENKEL F. 1913:
Römische Fingerringe der Rheinlande und der benachbarten Gebiete, Berlin.

HORVAT J. 2002:
The Hoard of Roman Republican Weapons from Grad near Šmihel. AVes 53, 117-192.

INTERCISA II 1957:
Grupa autora, Intercisa (Dunapentele) II. Geschichte der Stadt in der Römerzeit. AH XXXVI.

INTERCISA I 1976:
Die Gräberfeld von Intercisa I (E. B. Vago-I. Bona). Akademia Kiado, Budapest.

ISINGS C. 1957:
Roman Glass from Dated Finds. Archaeologica Traiectina edita ab Academiae Rheno-Traiectinae. Instituto Archaeologico II. J. B. Wolters, Groningen – Djakarta.

KONDIĆ J. 1994:
Kasnoantičko srebro, in: Antičko srebro u Srbiji (I. Popović), Narodni muzej u Beogradu, Beograd, 55-67.

KOŠČEVIĆ R. 1980:
Antičke fibule s područja Siska. Odjel za arheologiju Centra za povijesne znanosti, Zagreb.

KOŠČEVIĆ R. 1990:
Olovni privjesci iz Siska / Lead Pendants from Sisak. Prilozi 7, 23-30.

KOŠČEVIĆ R. 1991:
Antička bronca iz Siska. Umjetničko-obrtna produkcija iz razdoblja rimskog carstva. Odjel za arheologiju Instituta za povijesne znanosti Sveučilišta u Zagrebu, Zagreb.

KOŠČEVIĆ R. 1991/1:
Pečatne kapsule iz Siska / Seal Boxes from Sisak. Prilozi 8, 25-36.

KOŠČEVIĆ R. 1995:
Finds and Metalwork Production, in: R.Koščević-R. Makjanić, Siscia – Pannonia Superior. BARIntSer 621.

KOŠČEVIĆ R. 1996:
Nekoliko primjeraka staklene bižuterije iz rimskog razdoblja / Einige Exemplare der Glassbijouterie aus der römischen Epoche. Prilozi 10 (1993), 81-93.

KOŠČEVIĆ R. 1997:
Metalna produkcija antičke Siscije. Der Herstellung von Metalgegenständen im antiken Siscia. Prilozi 11 – 12 (1994 – 1995), 41-62.

KOŠČEVIĆ R. 1999:
Merkurove statuete iz Siscije / Figurines of Mercury from Siscia. Prilozi 15 – 16. (1998 – 1999), 21-28.

KOŠČEVIĆ R. 2000/1:
Sitni koštani i brončani predmeti iz Siscije / Small Bone and Bronze Objects from Siscia. Prilozi 17, 17-24.

KOŠČEVIĆ R. 2000/2:
Olovne pločice posebne namjene / Lead Tablets of Special Purpose. Prilozi 17, 95-101.

KOŠČEVIĆ R. 2000/3:
An additional review of seal boxes from the Roman period. Instrumentum 12, 14.

KOŠČEVIĆ R. 2001:
Daljnja opažanja o olovnim privjescima / Further Observations on Lead Pendants. Prilozi 18, 143-156.

KOŠČEVIĆ R. 2001/1:
Sitni metalni predmeti iz Siscije / Small metal artifacts from Siscia. Prilozi 18, 135-142.

KOŠČEVIĆ R. 2002:
Nekoliko rjeđih neobjavljenih nalaza iz Siscije / Some Rare Unpublished Finds from Siscia. Prilozi 19, 101-110.

KOŠČEVIĆ R. 2003:
Antičke staklene posude iz Siscije / Antique Glass Vessels from Siscia. Prilozi 20, 89-93.

KÜNZL E. 1984:
Das Gebet des Chryses (Homer, Ilias, 1. Gesang): Griechisches Epos und Römische Politik auf der vergoldeten Silberkanne des Octavius Menodorus. Mit Beiträgen von E. Foltz und G. Drews. JRGZ 31, 365-384.

LARESE A. 2004:
Vetri antichi del Veneto. CCAVV 8.

LARESE A.-ZERBINATI F. 1998:
Vetri antichi di Raccolte Concordiesi e Polesane. CCAVV 4.

LAZAR I. 2003:
Rimsko steklo Slovenije. The Roman Glass of Slovenia. Opera 7.

LOLIĆ T. 2007:
Integrated Rehabilitation Project Plan. Survey of the Architectural and Archaeological Heritage (IRPP/SAAH)

MÓCSY A. 1956:
Ólom árucímkék Sisciából. FolArch VIII, 89 – 110.

MOOSLEITNER F. 1980:
Handwerk und Handel, in: Die Kelten in Mitteleuropa-Kultur, Kunst, Wirtschaft. Salzburger Landesausstellung im Keltenmuseum Hallein Österreich 1. Mai-30. Sept. 1980, Salzburg, 93-101.

NENADIĆ V. 1987:
Prilog proučavanju antičke Sisciae. Prilozi 3 – 4 (1986 – 1987), 71-102.

NEVIODUNUM 1978:
Petru S. - Petru P., Neviodunum (Drnovo pri Krškem). KatMonLj 15.

POETOVIO I, II 1999:
Istenič J., Poetovio, Zahodna grobišča I, II (Katalog) – Grobne celote iz Deželnega muzeja Joanneuma v Gradcu. Poetovio, The Western Cemeteries I, II (Catalogue) – Grave-Groups in the Landesmuseum Johanneum, Graz. KatMonLj 32.

PREGLED 2000:
Pregled zaštitnih arheoloških istraživanja 1990-2000 – catalogue (Z. Burkowsky), Gradski muzej Sisak, Sisak.

RAVAGNAN G. L. 1994:
Vetri antichi del Museo Vetrario di Murano. Collezioni dello Stato. CCAVV 1.

RADMAN LIVAJA I. 2004:
Militaria Sisciensia – Nalazi rimske vojne opreme iz Siska u fundusu Arheološkog muzeja u Zagrebu. KatMonZ vol. 1, sv. 1.

RADNOTI A. 1938:
Bronzegefässe von Pannonien. DissPann II, 6.

ROTTLOFF A. 2000:
Römisches Glas, in: Die Römer zwischen Alpen und Nordmeer- Zivilisatorisches Erbe einer europäischen Militärmacht. Katalog-Handbuch zur Landesausstellung des Freistaates Bayern, Rosenheim 2000 (L. Wamser), Mainz, 133-138.

SENA CHIESA G. 1966:
Gemme del Museo Nazionale di Aquileia, Padova.

SISCIA 1995:
Koščević R., Finds and Metalwork Production, in: Siscia-Pannonia Superior, BAR IntSer 621, 1-38, Pl. 1-55.

STRENA BULICIANA 1924:
Strena Buliciana – Bulićev zbornik, Zagreb – Split.

ŠAŠEL J. 1974:
Siscia. PWRE Suppl. XIV, 702-741.

ŠAŠEL J. 1992:
Die Limes-Entwicklung in Illyricum. Situla 30 (Jaroslav Šašel: Opera selecta), 397-403.

ŠEPER M. 1954:
Jedan nalaz keramike iz Siska. AVes V, 2, 305-315.

TEKIJA 2004:
Cermanović Kuzmanović A. – Jovanović A., Tekija. Đerdapske sveske, Posebna izdanja 4, Arheološki institut-Narodni muzej-Centar za arheološka istraživanja, Beograd

TONIOLO A. 2000:
Vetri antichi del Museo Archeologico Nazionale di Este. CCAVV 6.

TRAGOVI 2003:
Na tragovima vremena. Iz arheološke zbirke Mateja Pavletića – katalog (Ed. A. Rendić-Miočević). Arheološki muzej u Zagrebu, Zagreb.

TRANSPARÊNCIAS 1998:
Transparências imperiais – Vidros romanos da Croácia – katalog (I. Fadić), Milão – Roma.

ULBERT G. 1959:
Die römische Donau-Kastelle Aislingen und Burghöfe. Limesforsch 1.

ULBERT G. 1969:
Das frührömische Kastell Rheingönheim. Limesforsch 9.

ULBERT Th. 1981:
Ad Pirum (Hrušica)-spätrömische Passbefestigung in der Julischen Alpen. Der Deutsche Beitrag zu den Slowenisch-Deutschen Grabungen 1971 – 1973. VeröffKommErf 1981.

UNZ C. 1972:
Römische Militarfunde aus Baden – Aquae Helveticae. JahresberGPV 1971, 41-58

UNZ C. 1974:
Römische Funde aus Windisch im ehemaligen Kantonalen Antiquarium Araau. Jahresber 1973, 11-42.

VERMEULE C. C. 1974:
The goddess Roma in the art of the Roman empire. Cambridge (Mass).

VIDOŠEVIĆ I. 2003:
Rimska keramika s lokaliteta Starčevićeve ulice 37 u Sisku / Roman pottery from the site at 37 Starčević street in Sisak.

Godišnjak Gradskog muzeja Sisak III – IV (2002 – 2003), Sisak, 11-74.

VIKIĆ BELANČIĆ B. 1971:
Antičke svjetiljke u Arheološkom muzeju u Zagrebu. VAMZ 3. ser., sv. V, 97 – 182.

VIKIĆ BELANČIĆ B. 1976:
Keramika grublje fakture u južnoj Panoniji s osobitim obzirom na urne i lonce. Aves XXVI (1975 – 1976), 74-128.

VRBANOVIĆ S. 1981:
Prilog proučavanju topografije Siscije. Izdanja HAD 6, 187-200.

WALKE N. 1965:
Das römische Donaukastell Straubing-Sorviodurum. Limesforsch 3.

WIEWEGH Z. 2002:
Jugoistočna nekropola Siscije – catalogue, Gradski muzej Sisak, Sisak.

WIEWEGH Z. 2003:
Rimske prstenaste fibule iz Antičke zbirke Gradskog muzeja Sisak, Godišnjak GMS III – IV (2002 – 2003), 75-88.

ZAMPIERI G. 1998:
Vetri antichi del Museo Civico Archeologico di Padova. CCAVV 3.

Pl. 1

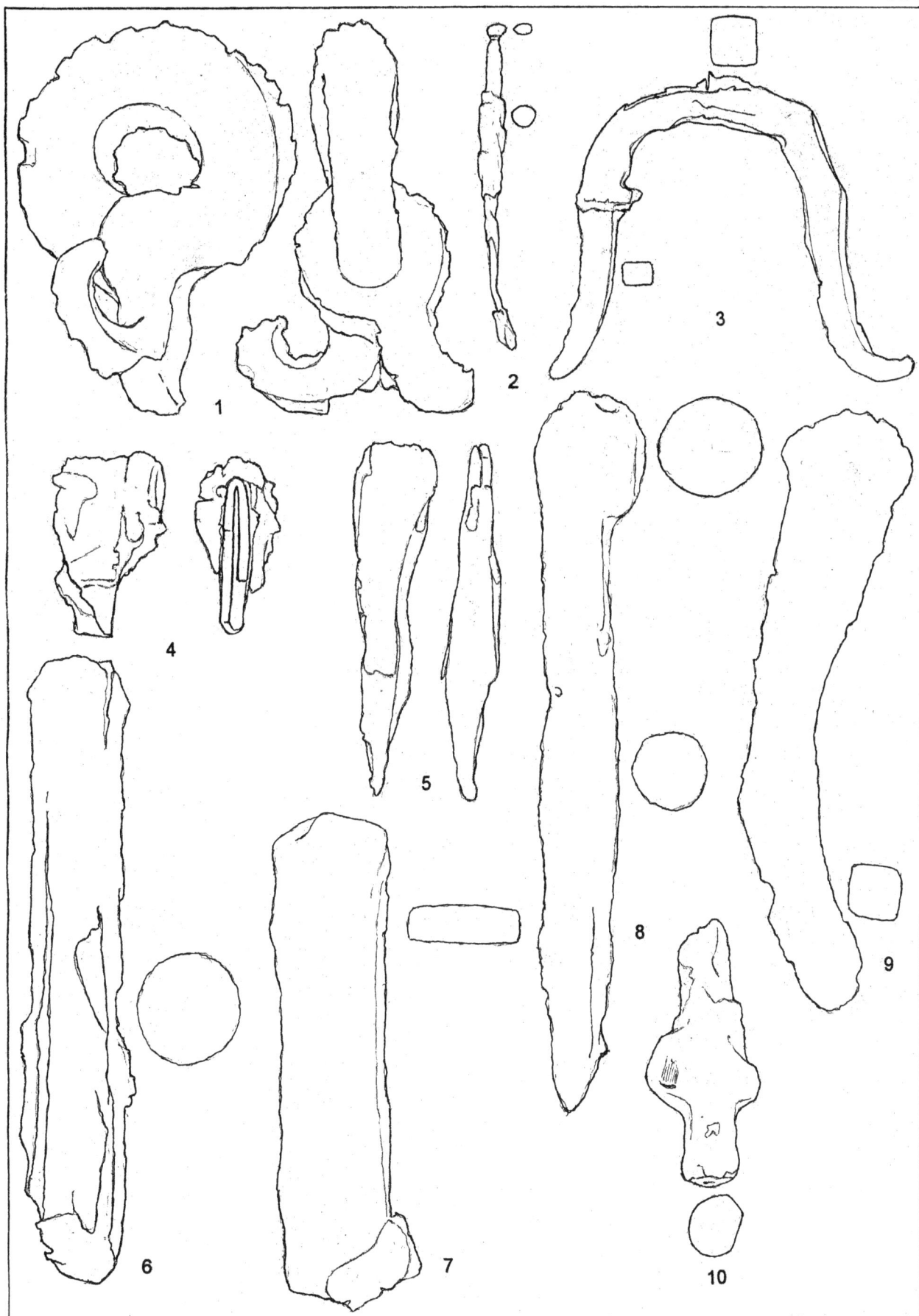

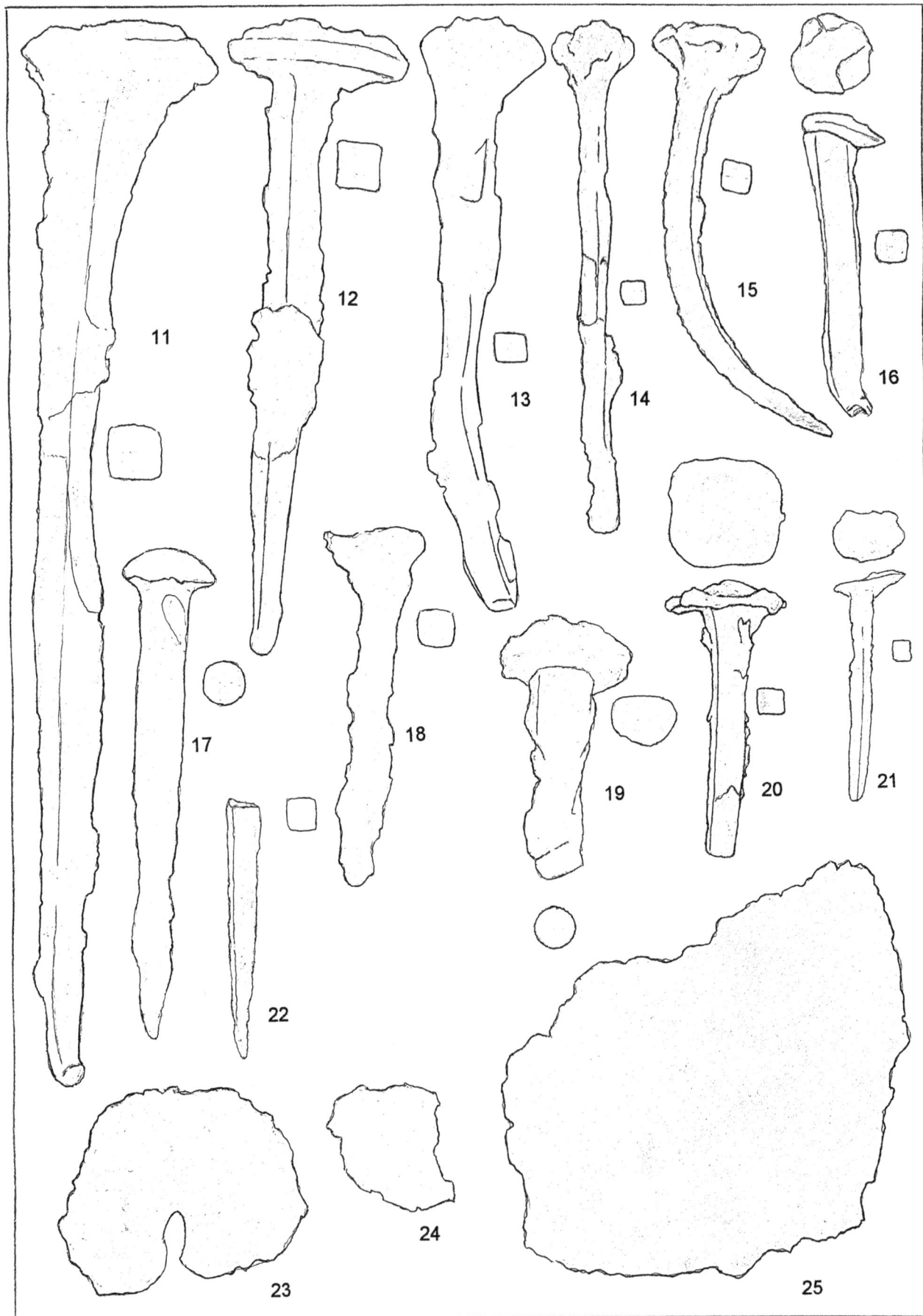

Pl. II

11

12

13

14

15

16

17

18

19

20

21

22

23

24

25

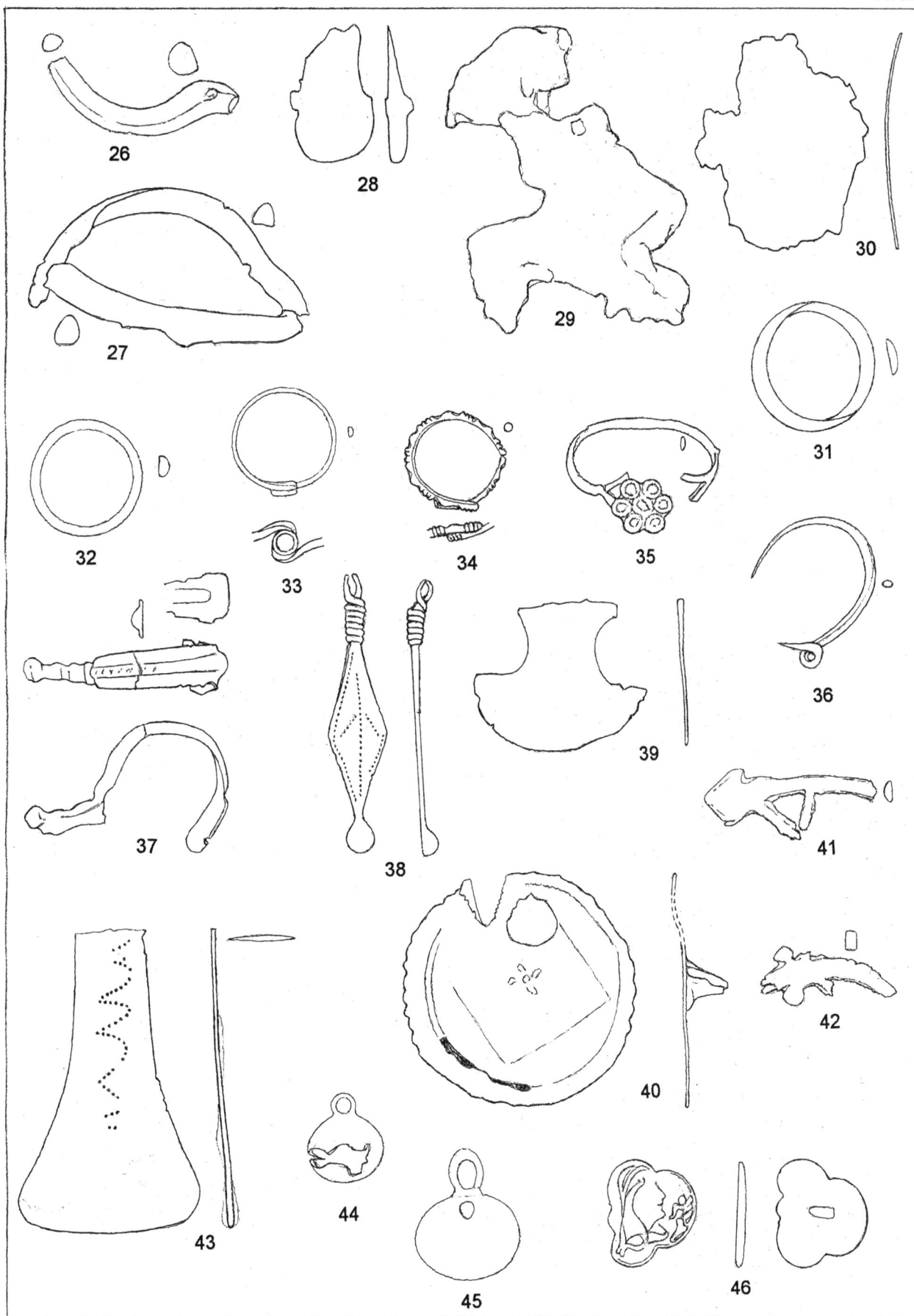

26

28

29

30

27

31

32

33

34

35

36

37

38

39

41

40

42

43

44

45

46

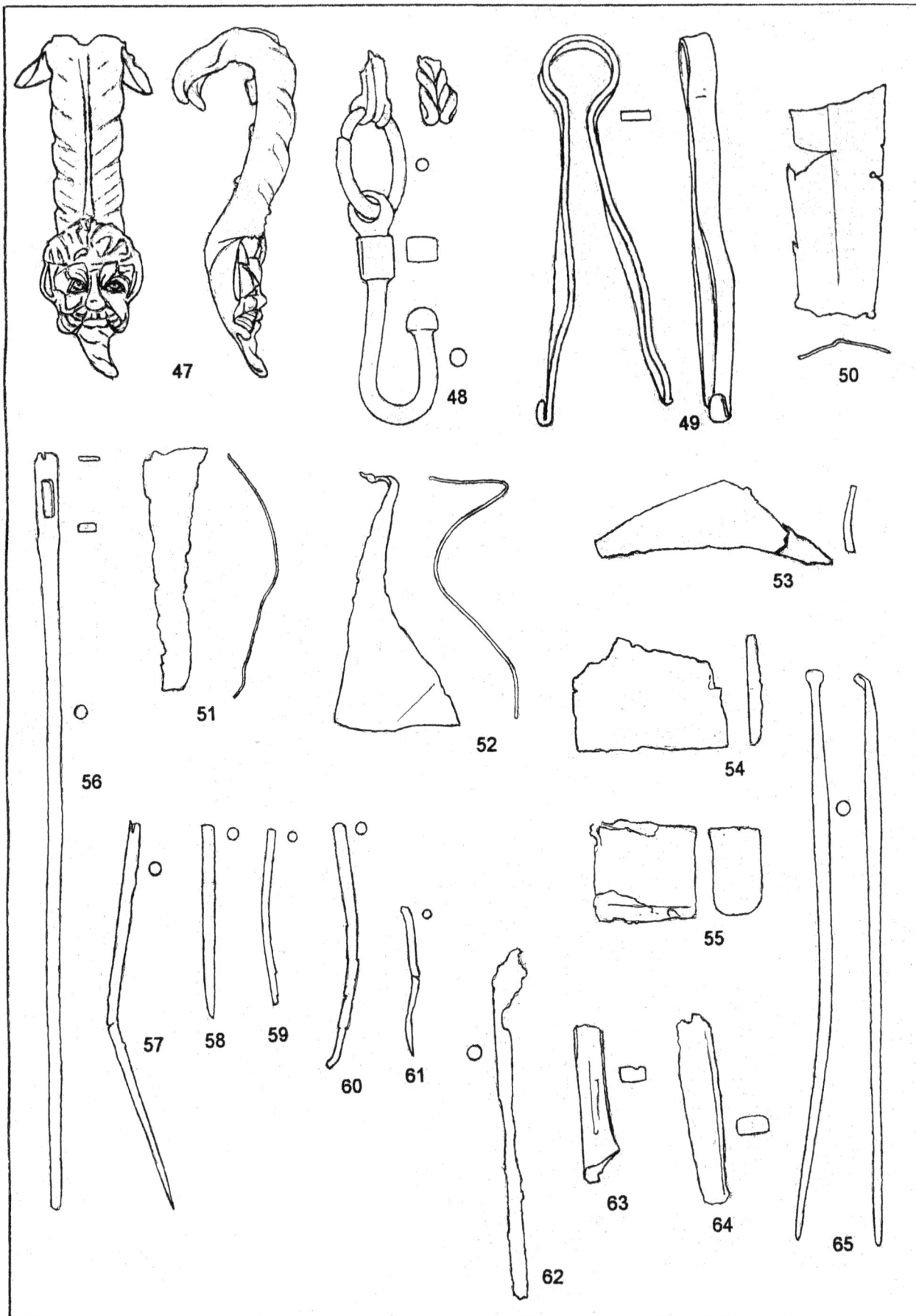

Pl. IV

47

48

49

50

51

52

53

54

55

56

57

58

59

60

61

62

63

64

65

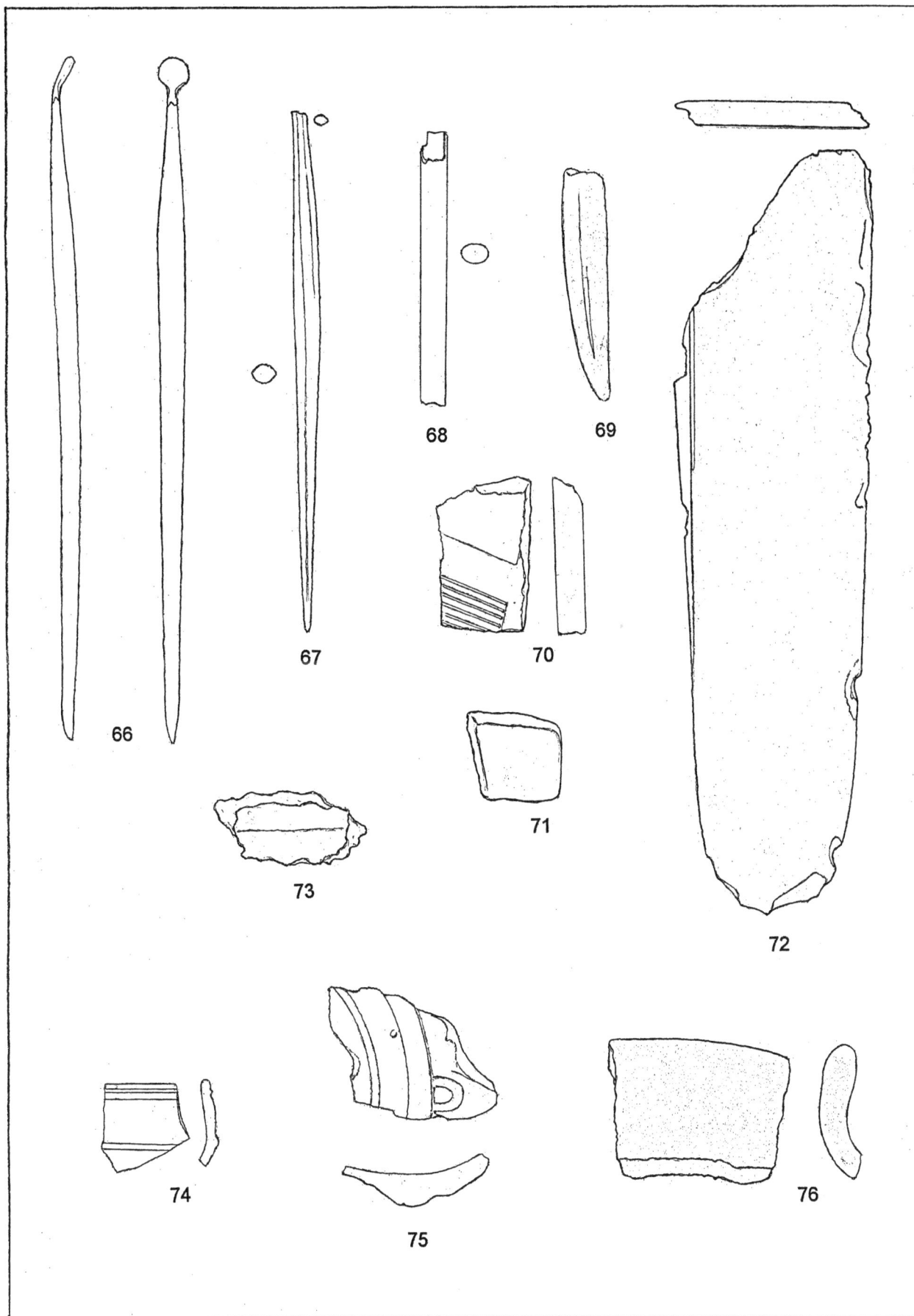

66

67

68

69

70

71

72

73

74

75

76

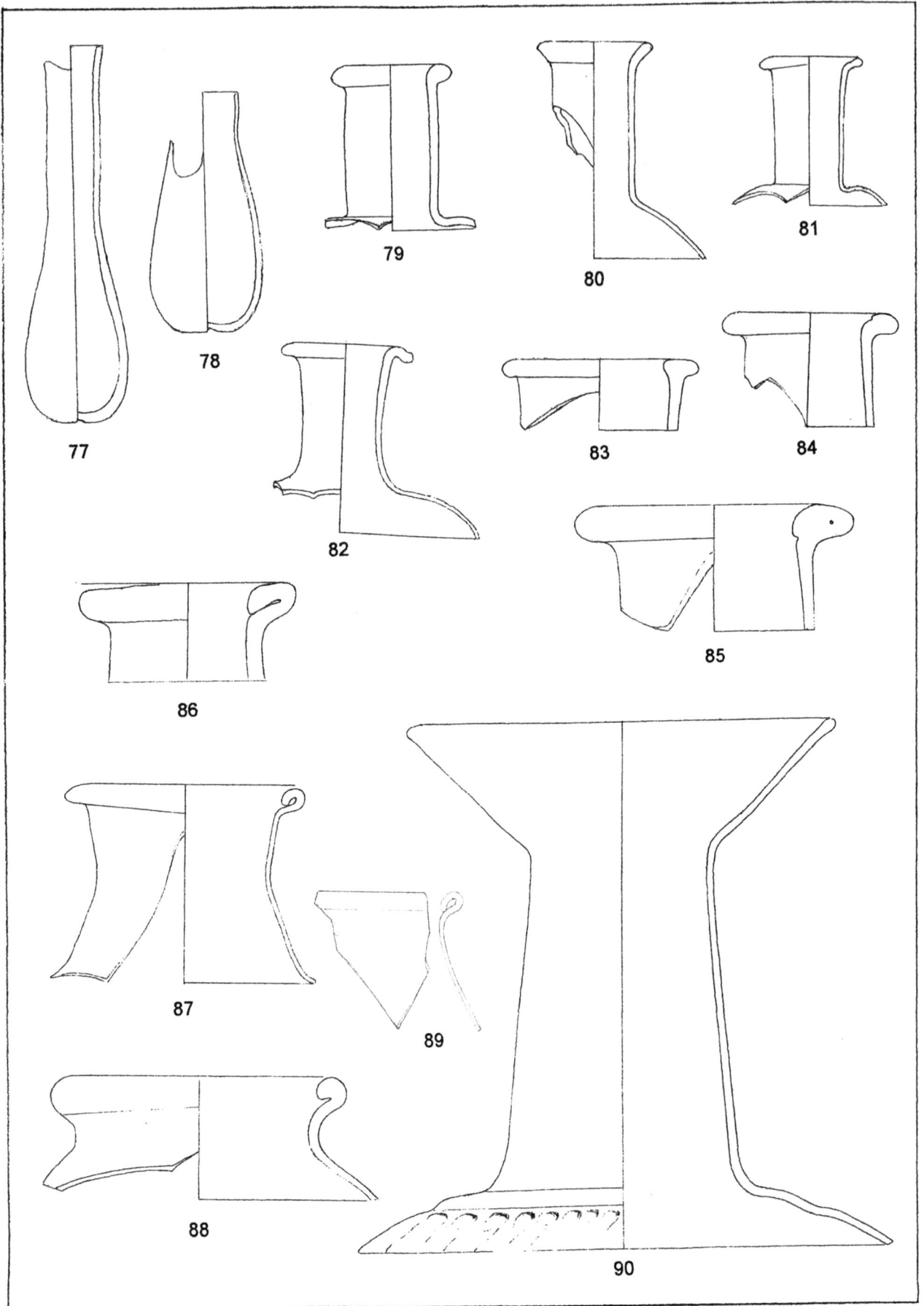

Pl. VI

77

78

79

80

81

82

83

84

85

86

87

88

89

90

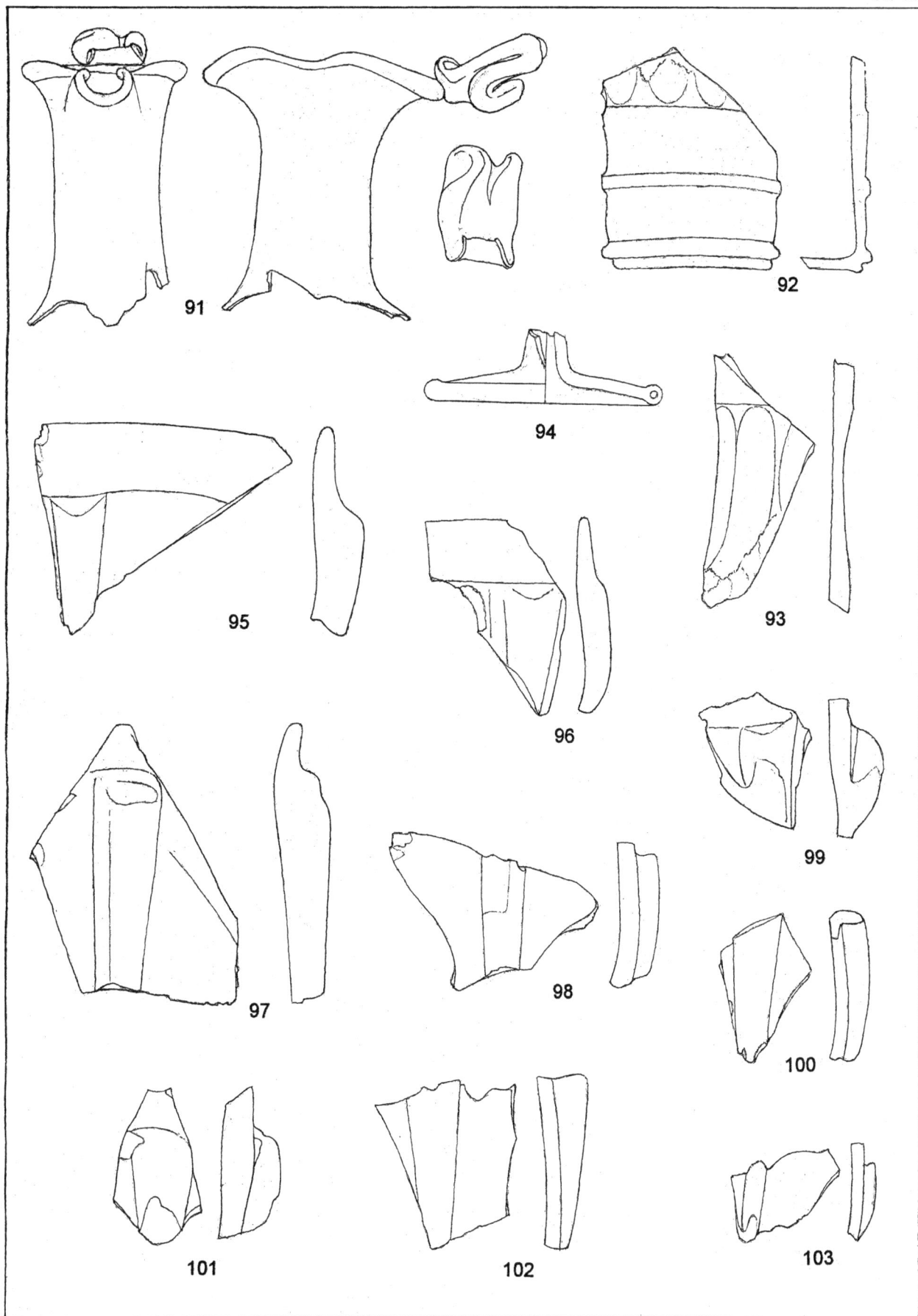

Pl. VII

91

92

94

95

96

93

97

98

99

100

101

102

103

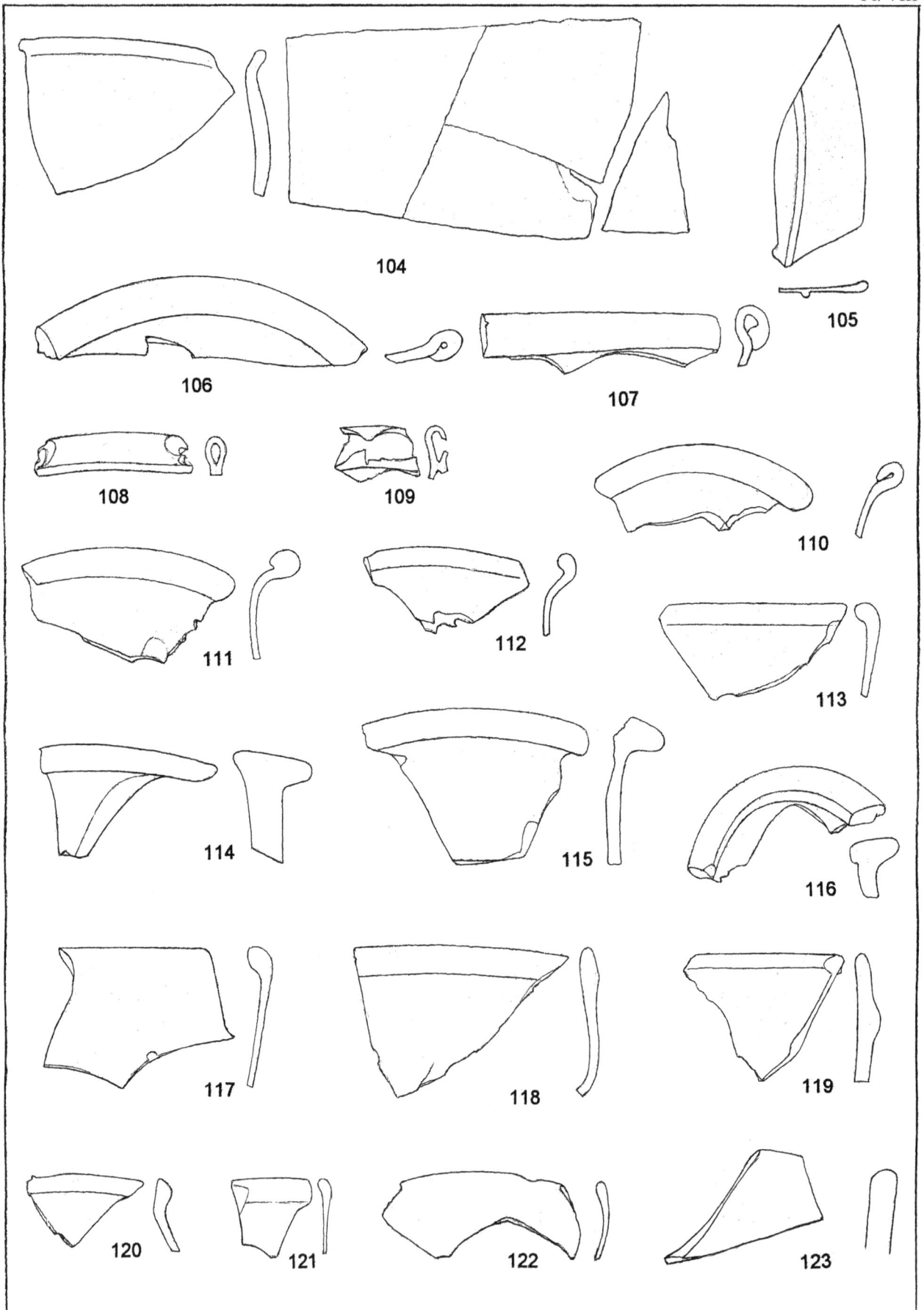

Pl. VIII

104

105

106

107

108

109

110

111

112

113

114

115

116

117

118

119

120

121

122

123

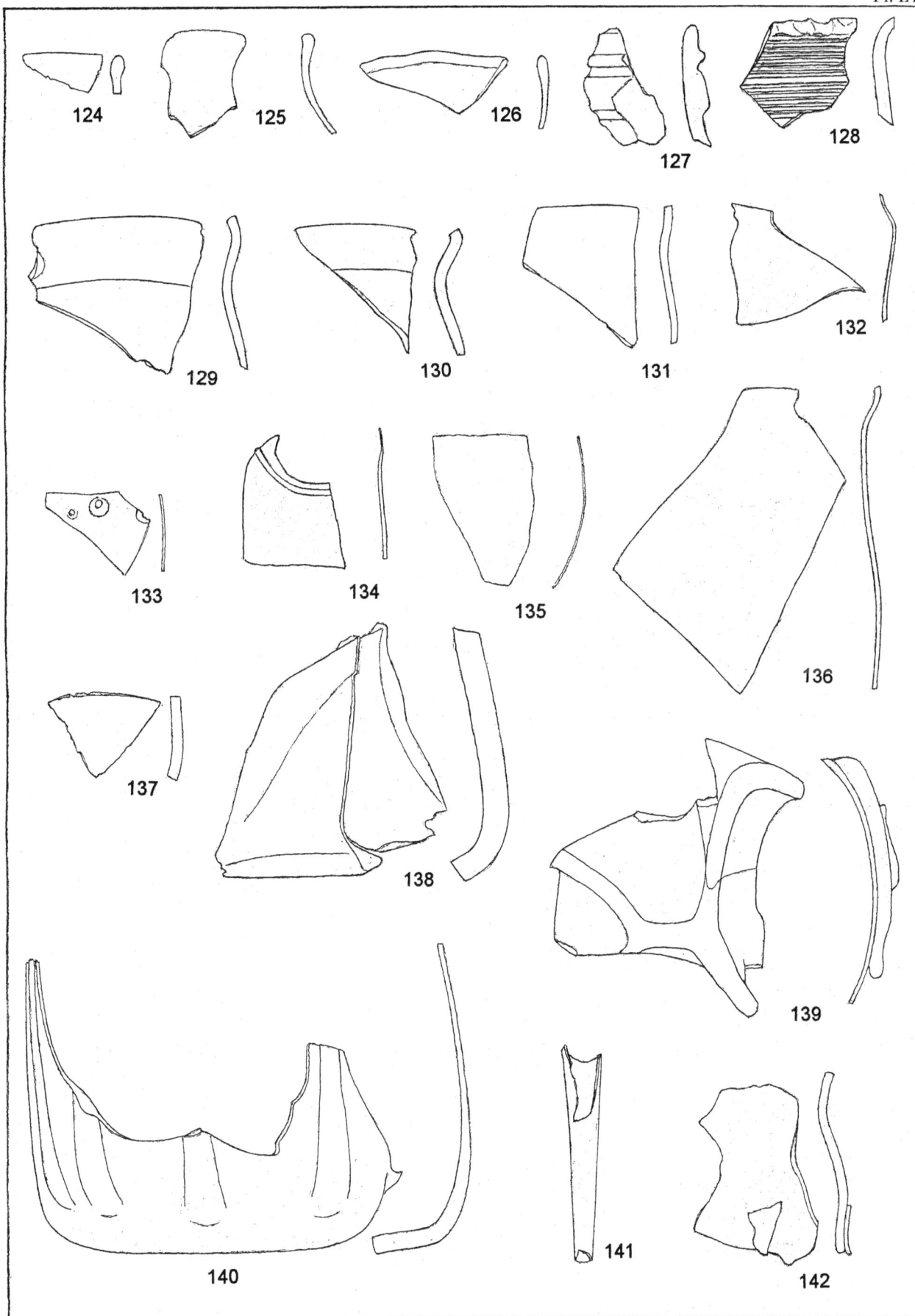

Pl. IX

124

125

126

127

128

129

130

131

132

133

134

135

136

137

138

139

140

141

142

Pl. X

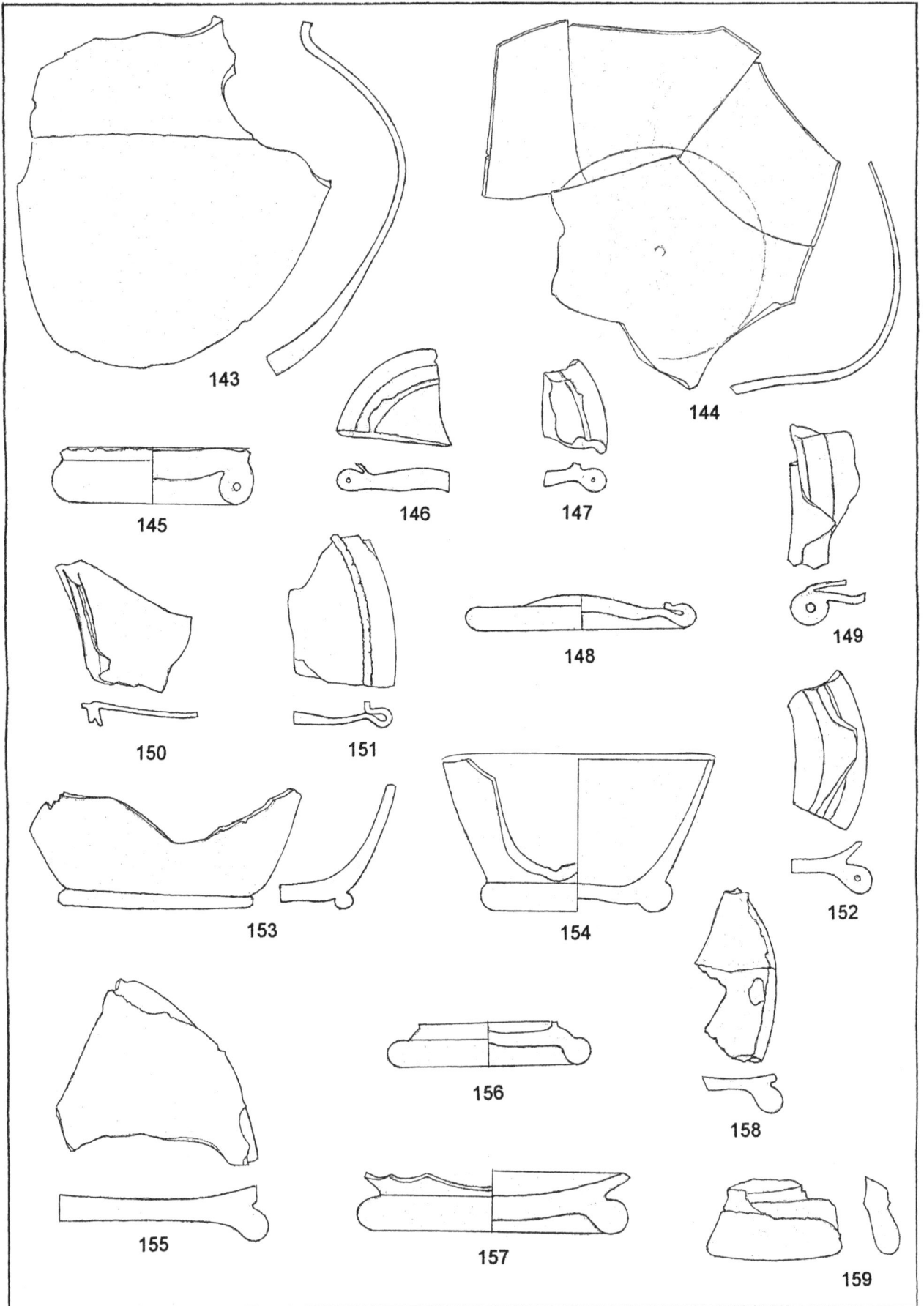

143

144

145

146

147

148

149

150

151

152

153

154

155

156

157

158

159

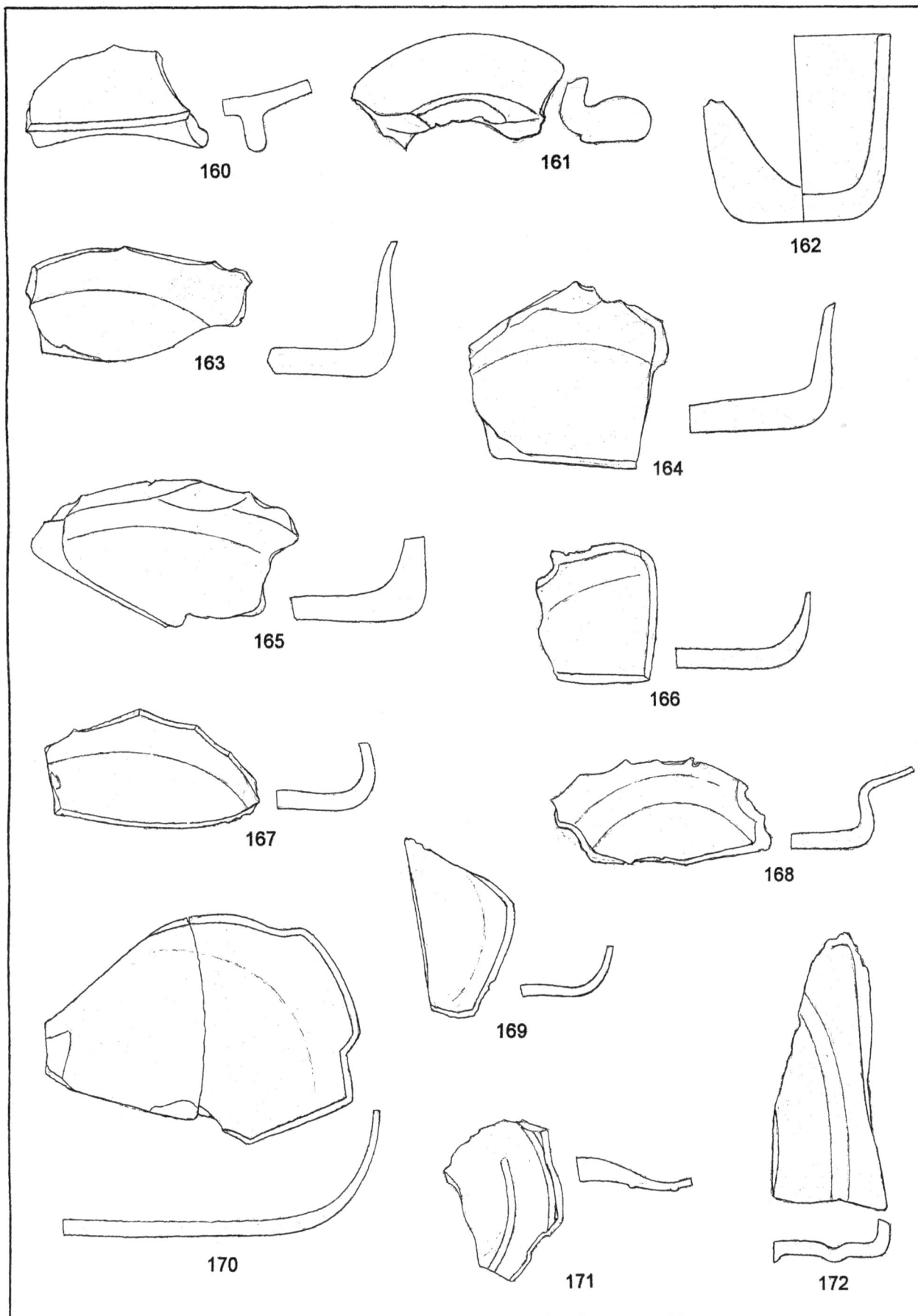

160

161

162

163

164

165

166

167

168

169

170

171

172

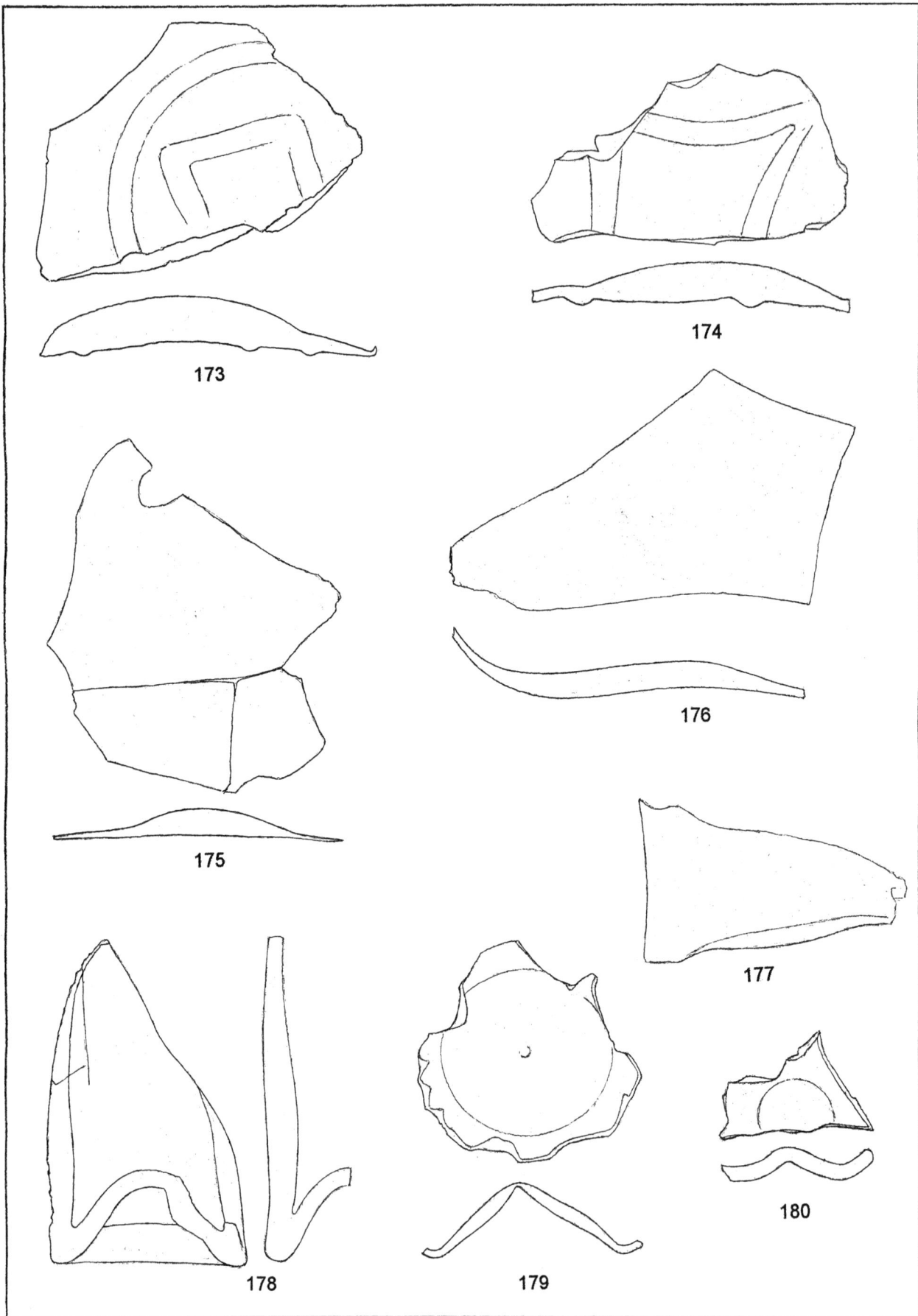

Pl. XII

173

174

175

176

177

178

179

180

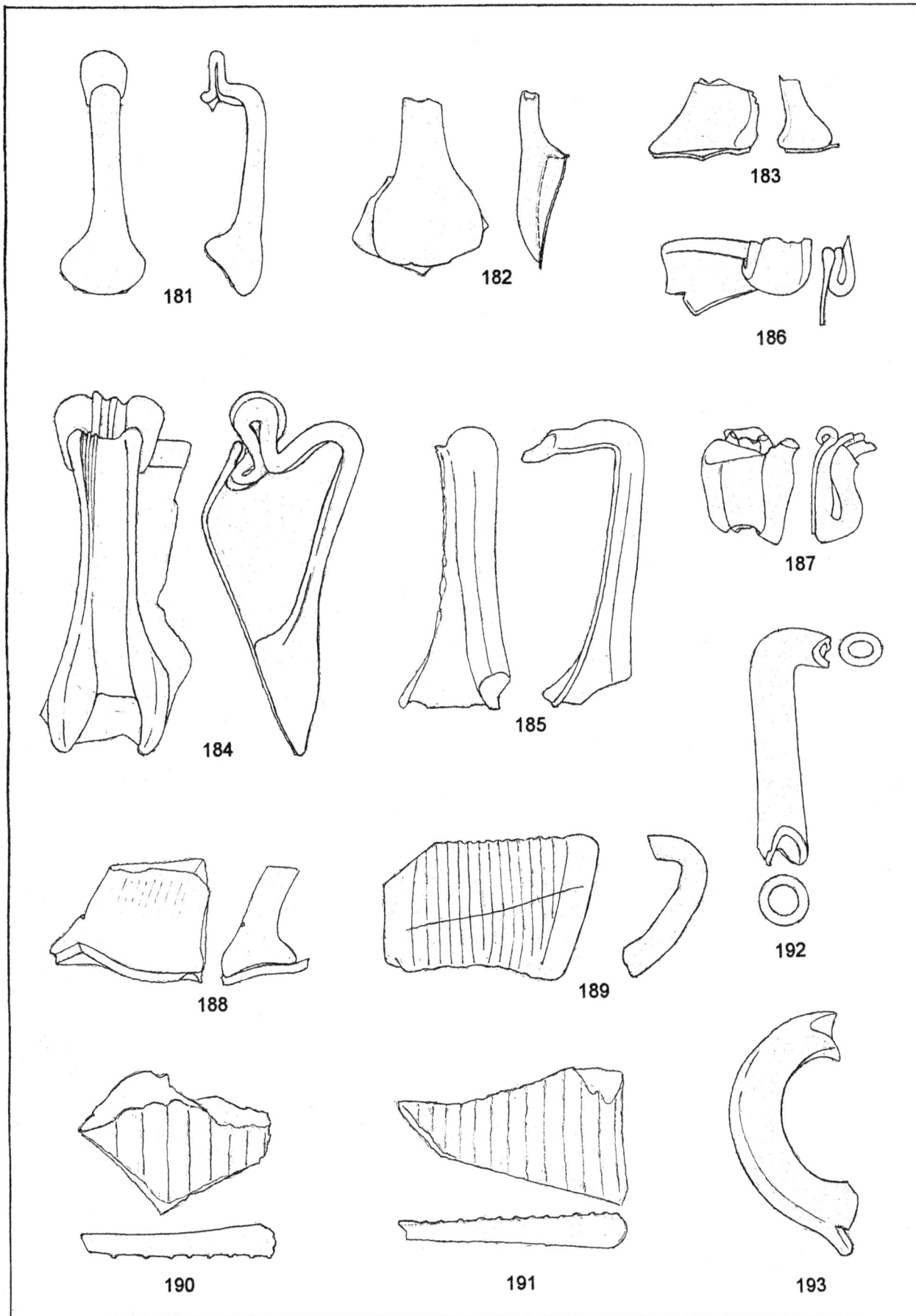

181

182

183

186

184

185

187

188

189

192

190

191

193

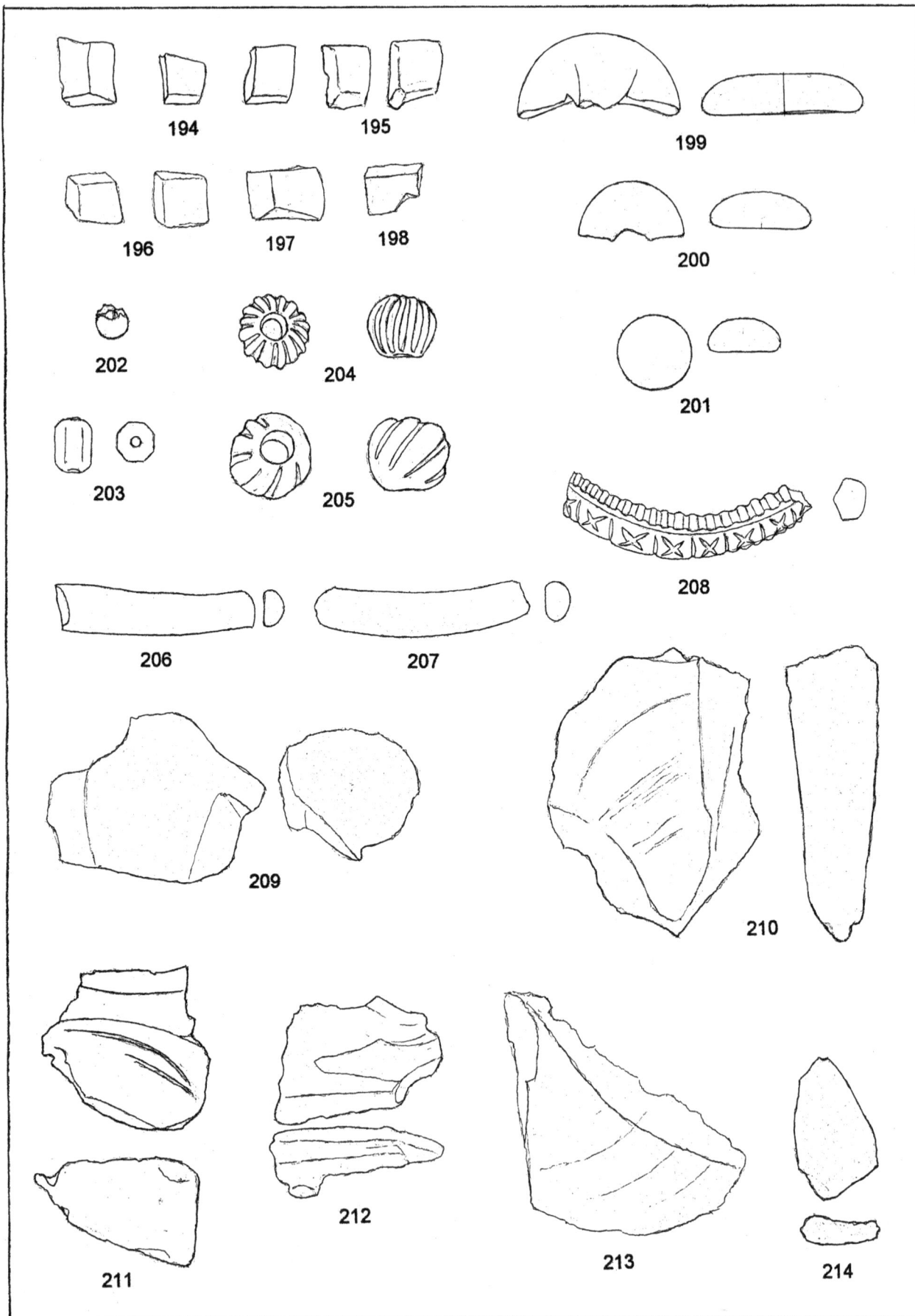

Pl. XIV

194

195

199

196 197 198

200

202 204

201

203 205

208

206 207

209

210

211 212 213 214

Iron. Fig. 1: double hook. Fig. 2: pin. Fig. 3: handle. Fig. 4: fragment of bent sheet metal. Figs. 5, 8, 9: spikes. Figs. 6, 7: fragments of bars. Fig. 10: point of an arrowhead. Figs. 11 - 15: nails.

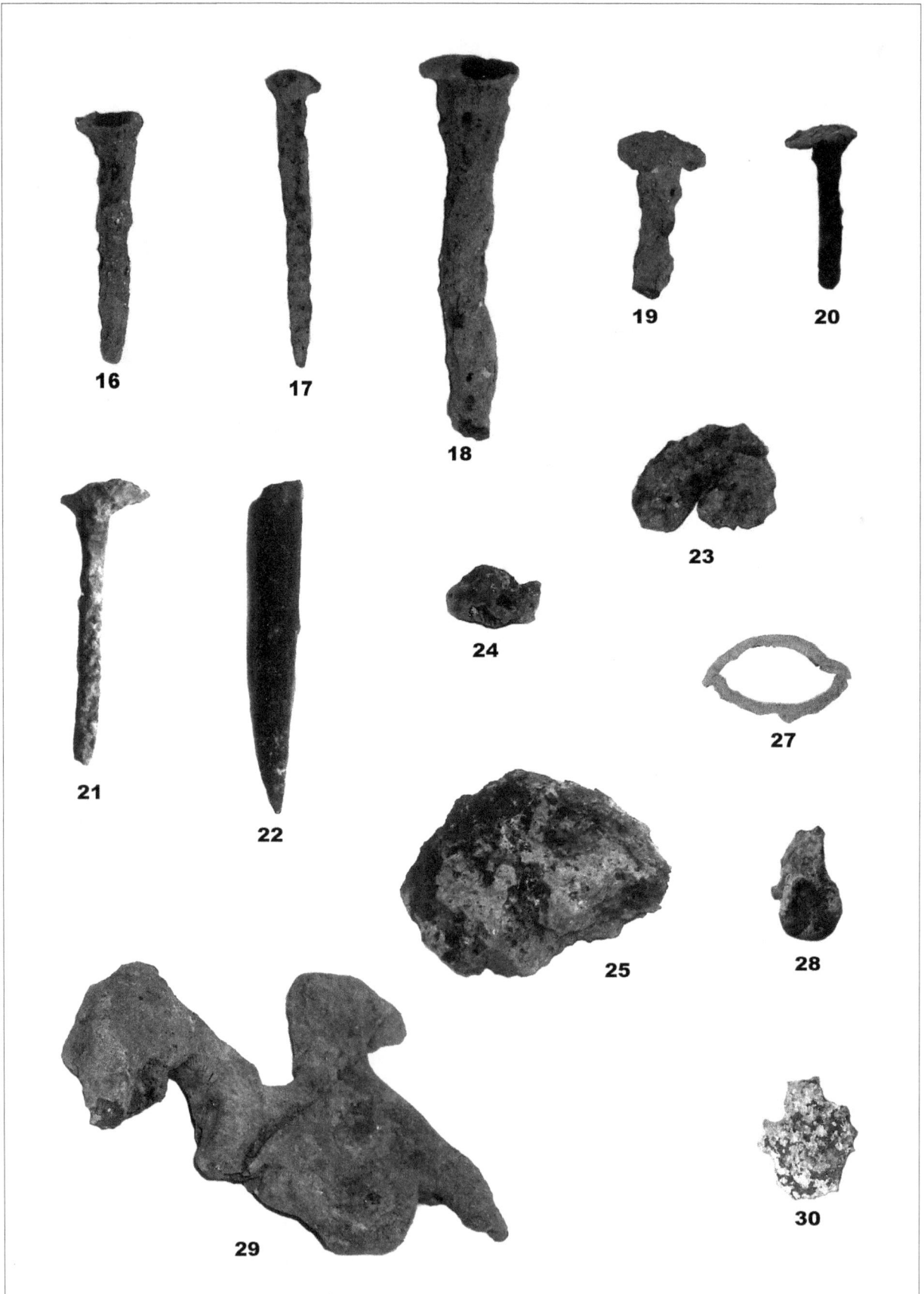

Iron. Figs. 16 - 21: nails. Fig. 22: shank of a nail. Fig. 23: clump of burnt iron. Figs. 24, 25: iron slag. White metal. Fig. 27: bent rod. Fig. 28: piece of metal. Fig. 29: piece of unworked lead. Fig. 30: piece of sheet metal.

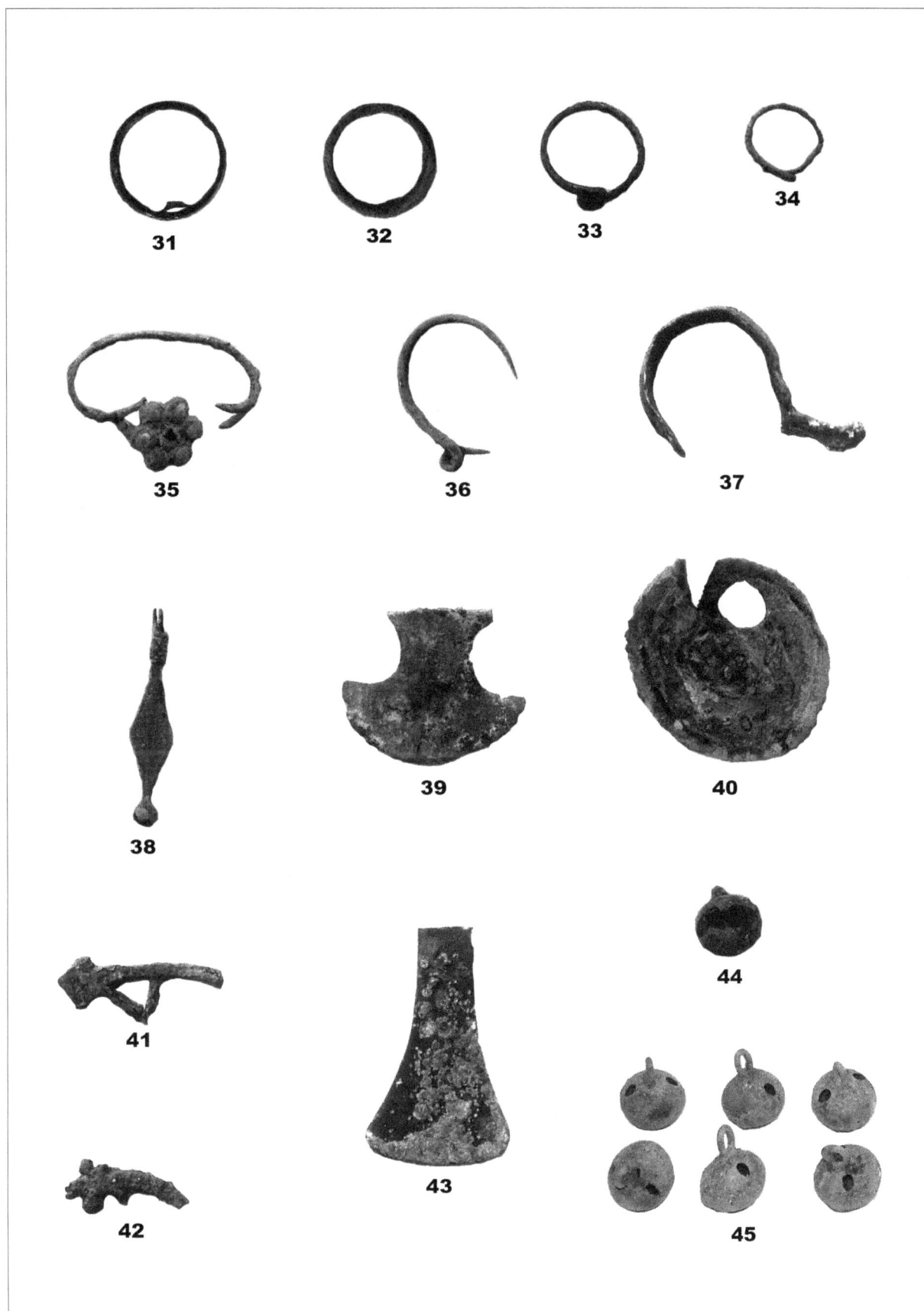

Silver (?). Fig. 31: ring. Bronze. Figs. 32 – 35: rings. Fig. 36: earring. Fig. 37: fibula. Fig. 38: pendant. Fig. 39: fragment of a larger object. Fig. 40: mount. Figs. 41, 42: mount fragments. Fig. 43: fragment of a larger object. Figs. 44, 45: buttons.

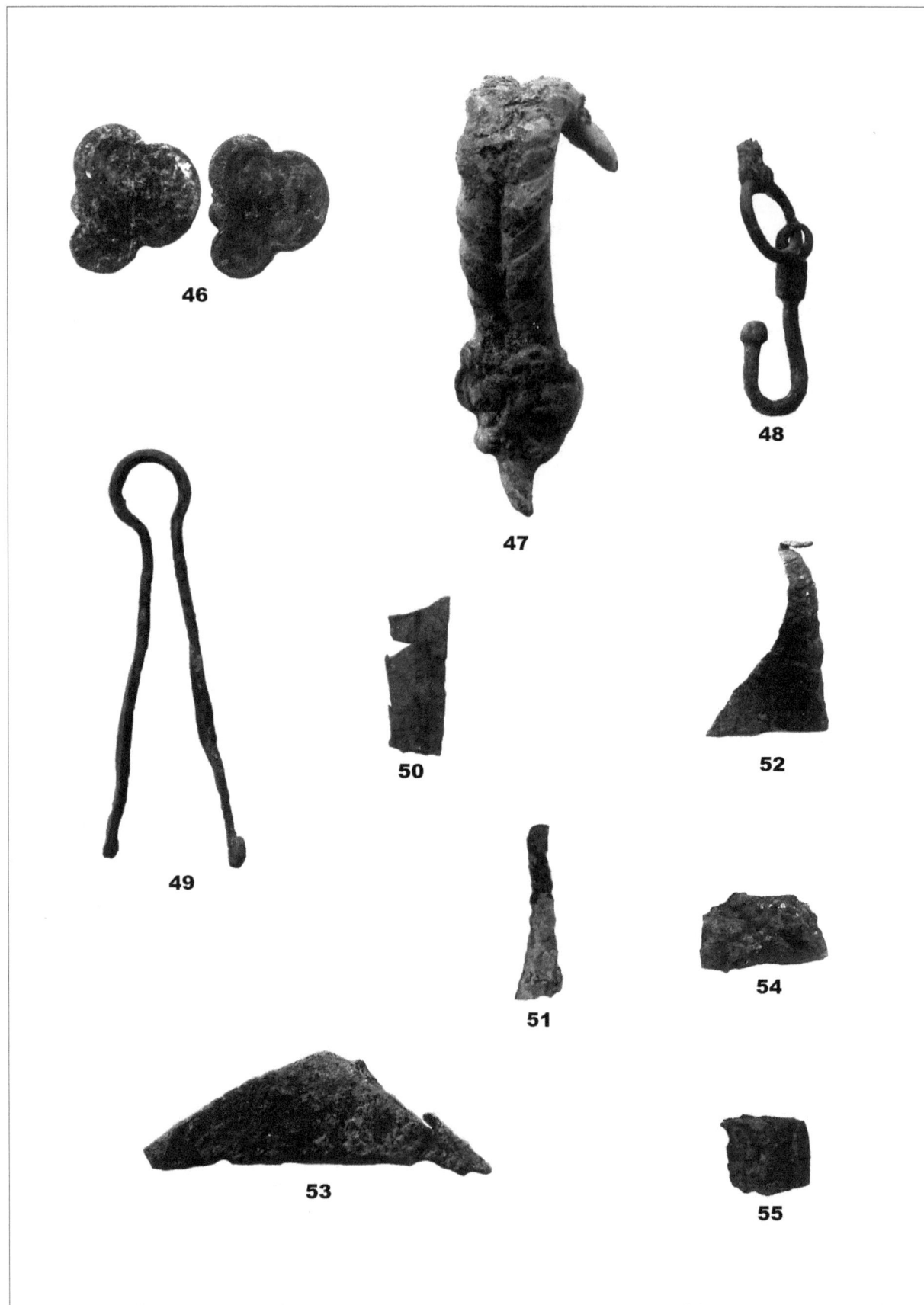

Bronze. Fig. 46: pair of appliqués. Fig. 47: handle. Fig. 48: hook. Fig. 49: double hook. Fig. 50: fragment of a larger object. Figs. 51 - 54: pieces of thick and thin sheet metal. Fig. 55: fragment of a thick rod.

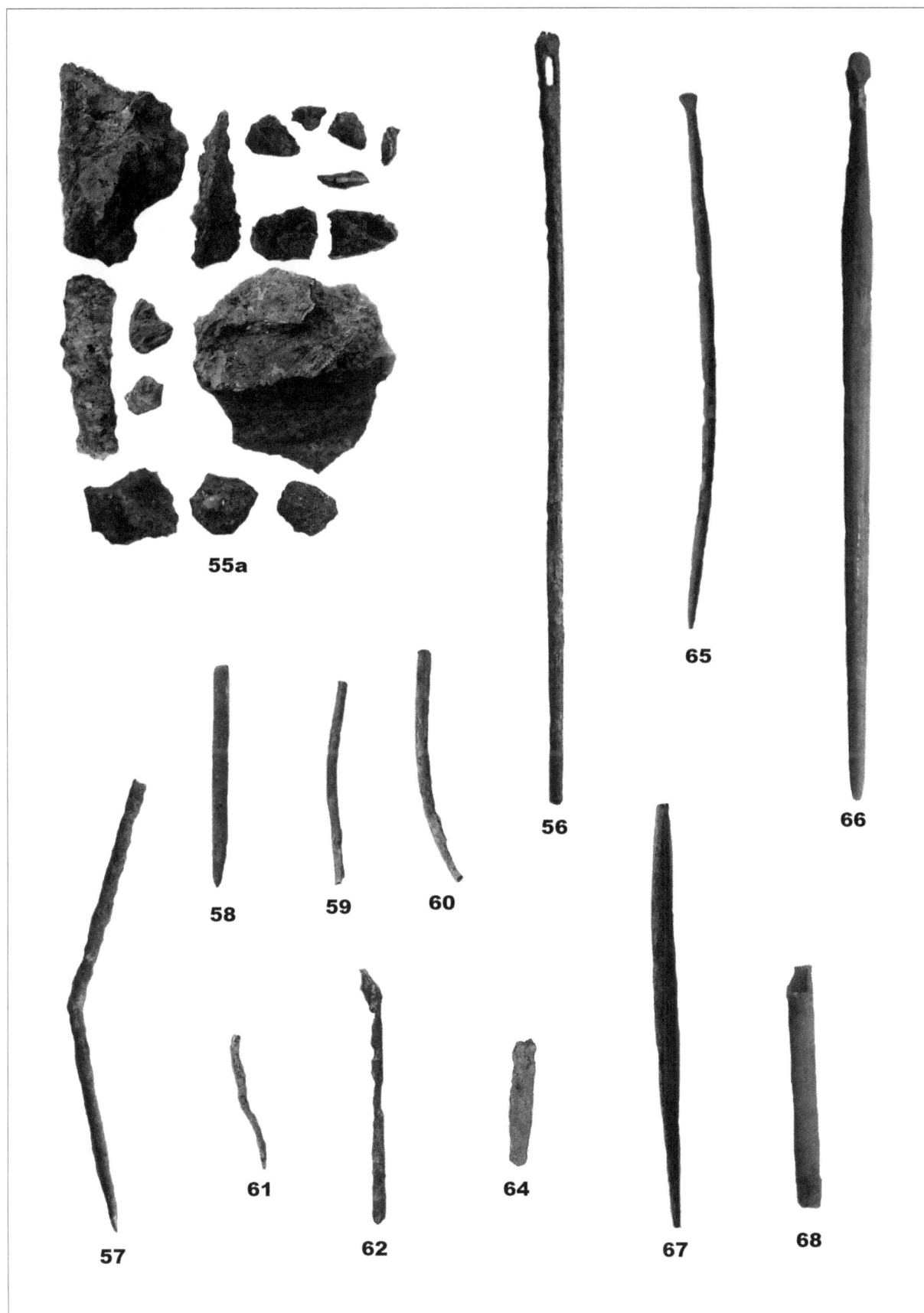

55a

56

65

66

58

59

60

57

61

62

64

67

68

Bronze. Fig. 55a: group of shapeless fragments under layers of patina. Figs. 56 - 62: pins and pin shanks of various sizes. Fig. 64: fragment of a bar. Fig. 65: probe. Bone.Fig. 66: probe. Fig. 67: holder for a probe or shank of a pin. Fig. 68: part of a small tube.

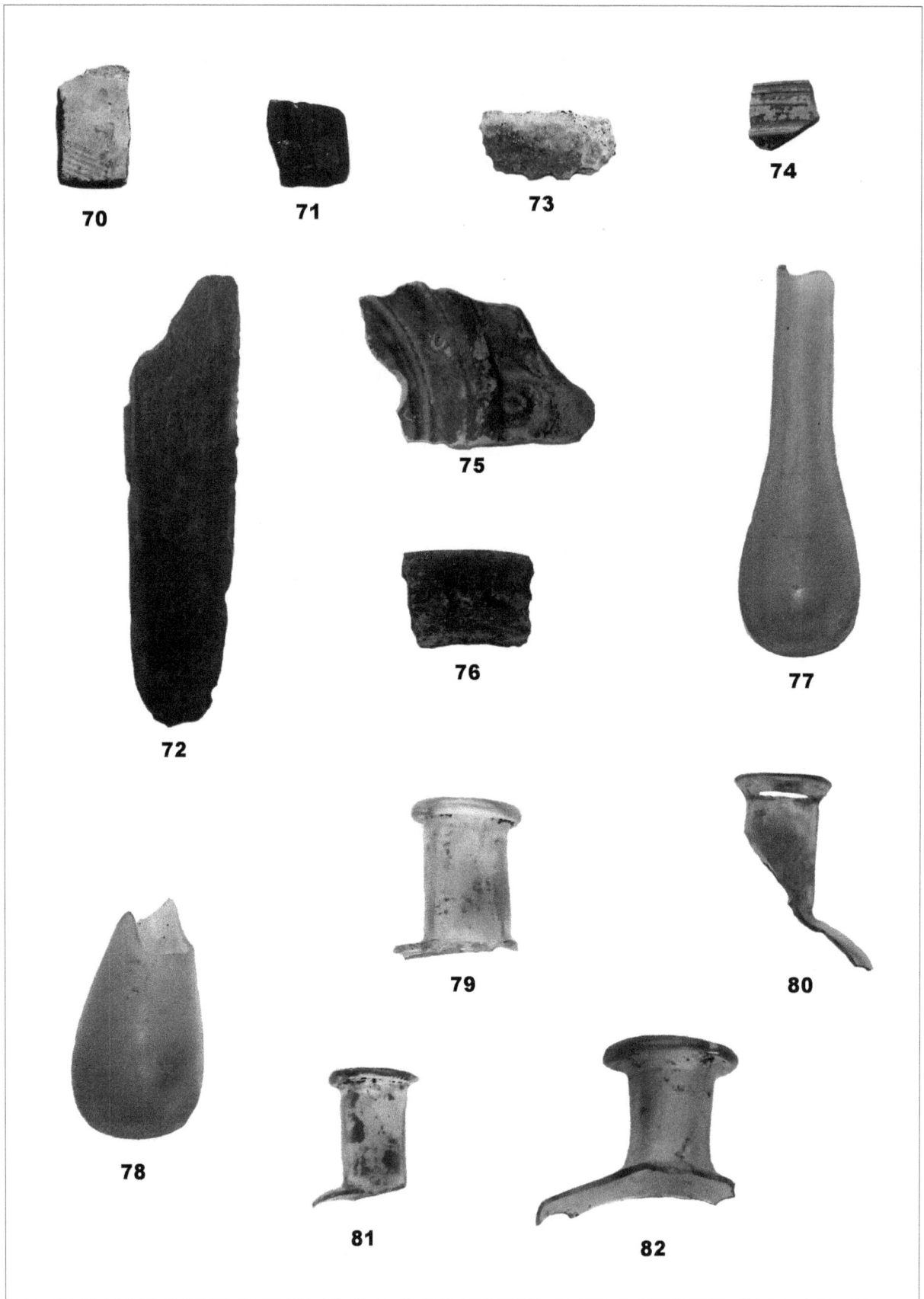

Stone. Fig. 70: fragmnent of a slab. Fig. 71: small cube. Fig. 72: whetstone. Plaster. Fig. 73: piece of wall painting. Pottery. Figs. 74, 76: fragments of vessels. Fig. 75: fragment of a clay lamp. Glass. Figs. 77, 78: balsamaria. Figs. 79 - 82: necks of small flasks.

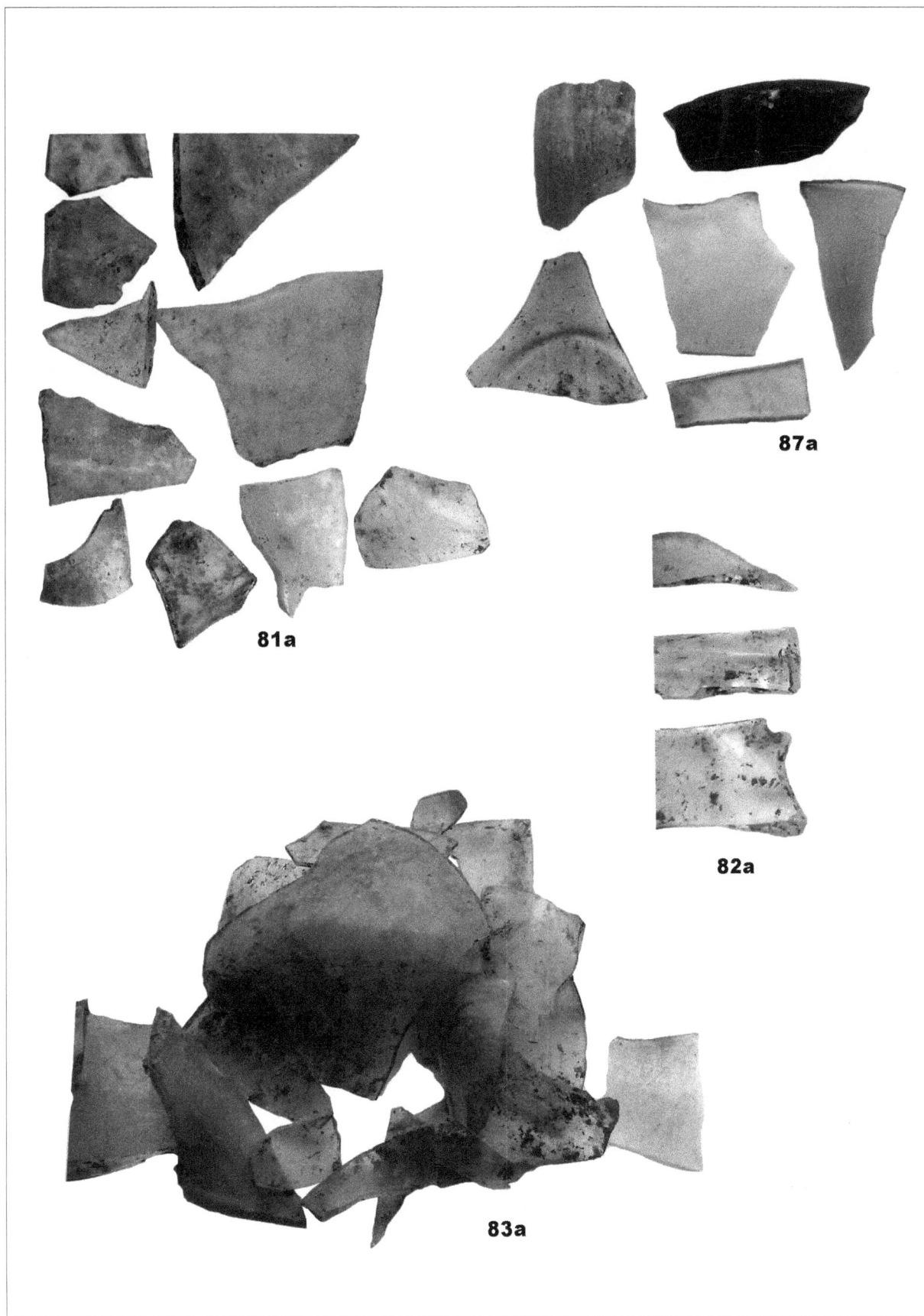

Glass. Figs. 81a - 83a, 87a: group of fragments with which items 81 - 83, 87 were found.

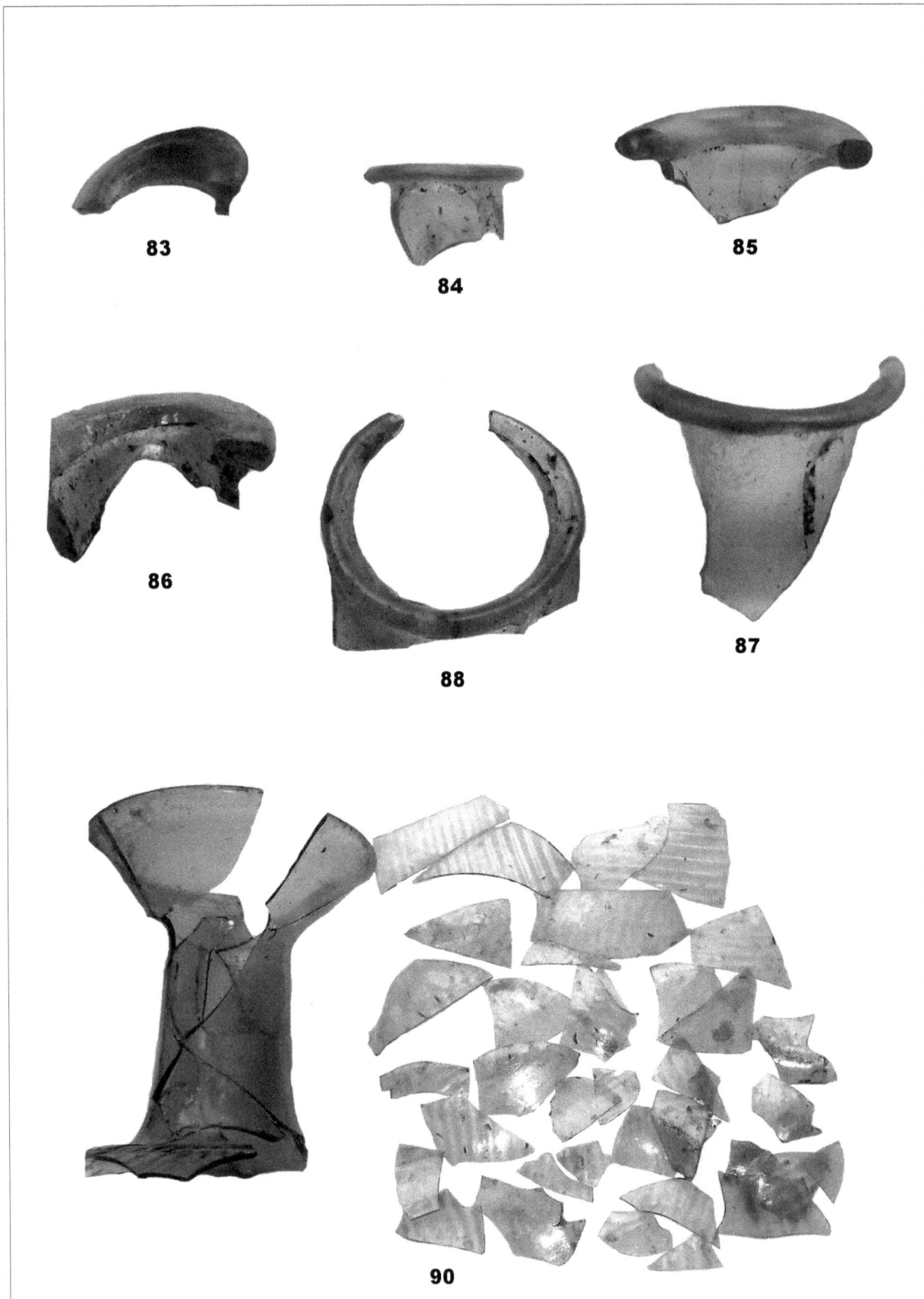

Glass. Figs. 83 - 88: fragments of the edges or necks or rims. Fig. 90: neck, rim, and wall fragments of a globular flask.

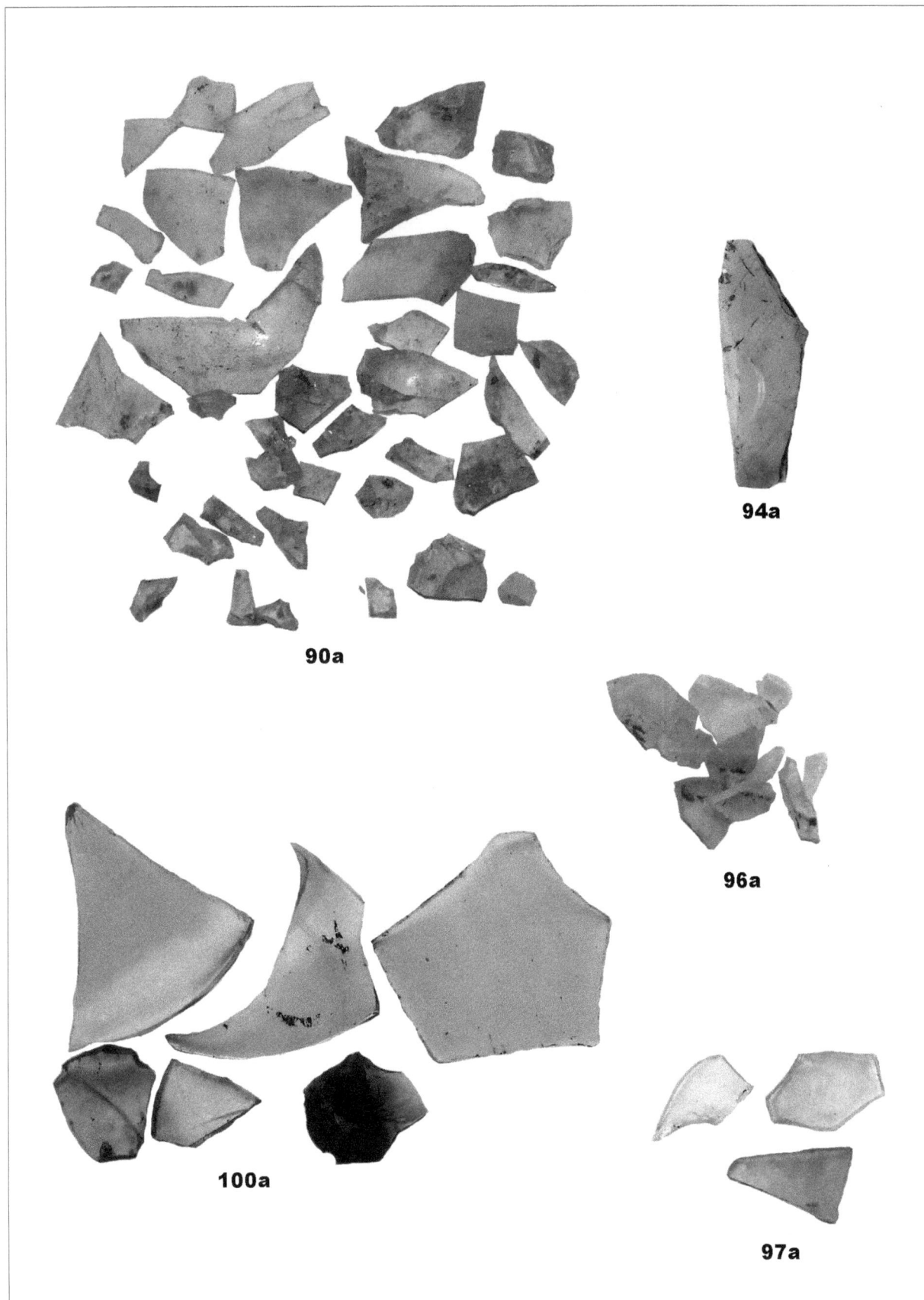

Glass. Figs. 90a, 94a, 96a, 97a, 100a: group of fragments with which items 90, 94, 96, 97, 100 were found.

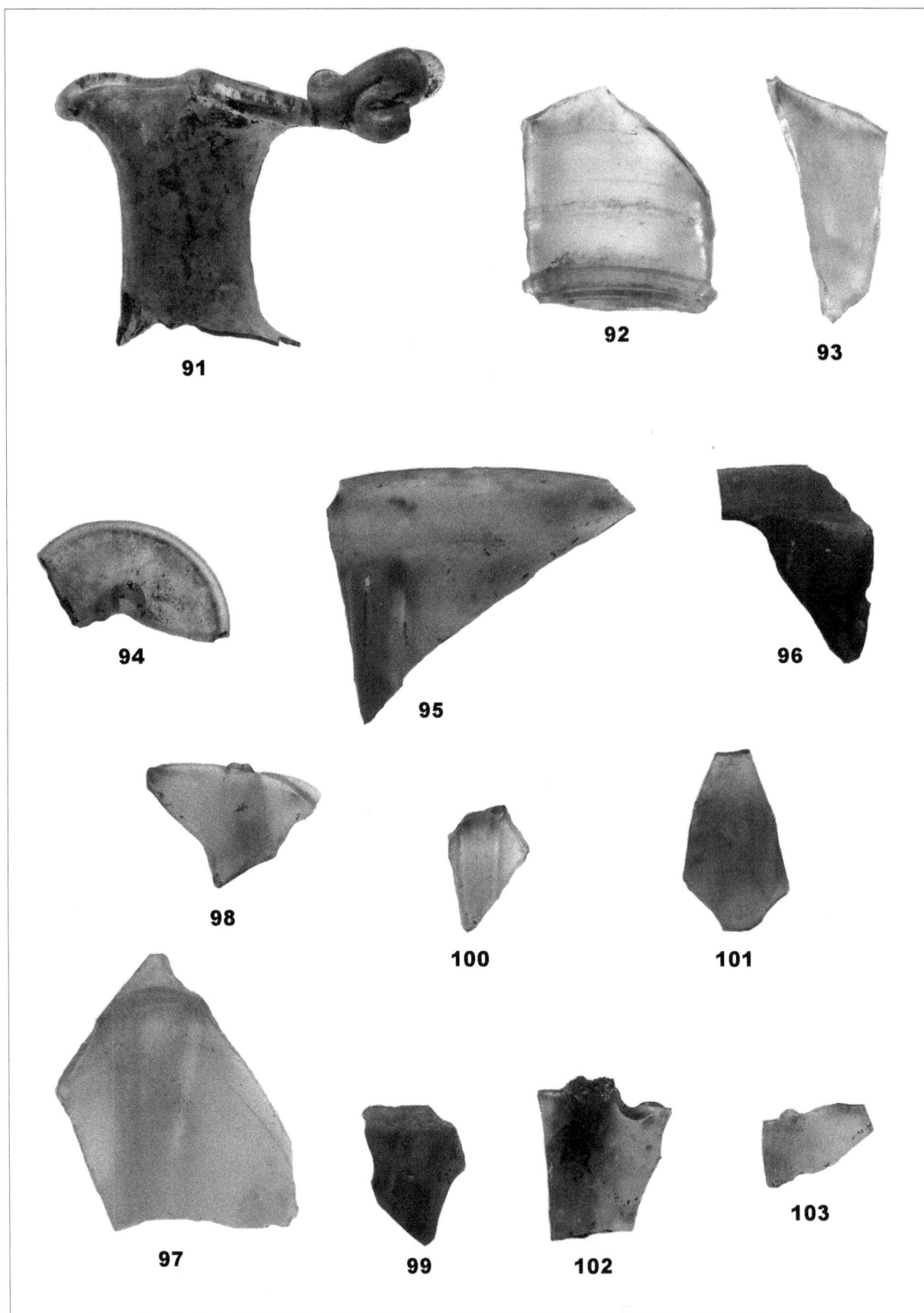

Glass. Fig. 91: neck of a small jug with a tri-lobed spout. Figs. 92, 93: beaker fragments. Fig. 94: part of the base of a goblet. Figs. 95 – 103: bowl fragments.

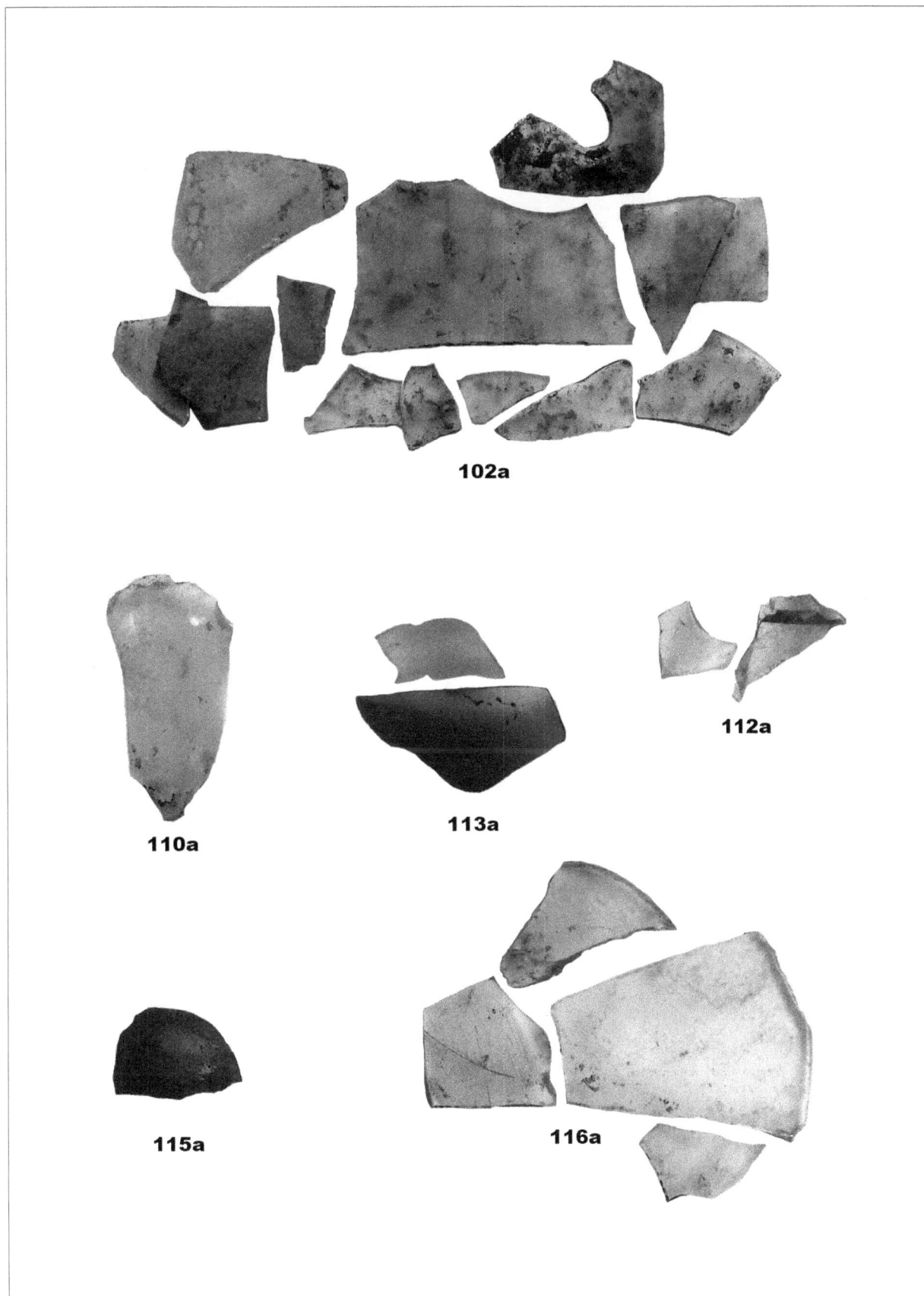

Glass. Figs. 102a, 110a, 112a, 113a, 115a, 116a: group of fragments with which items 102, 110, 112, 113, 115, 116 were found.

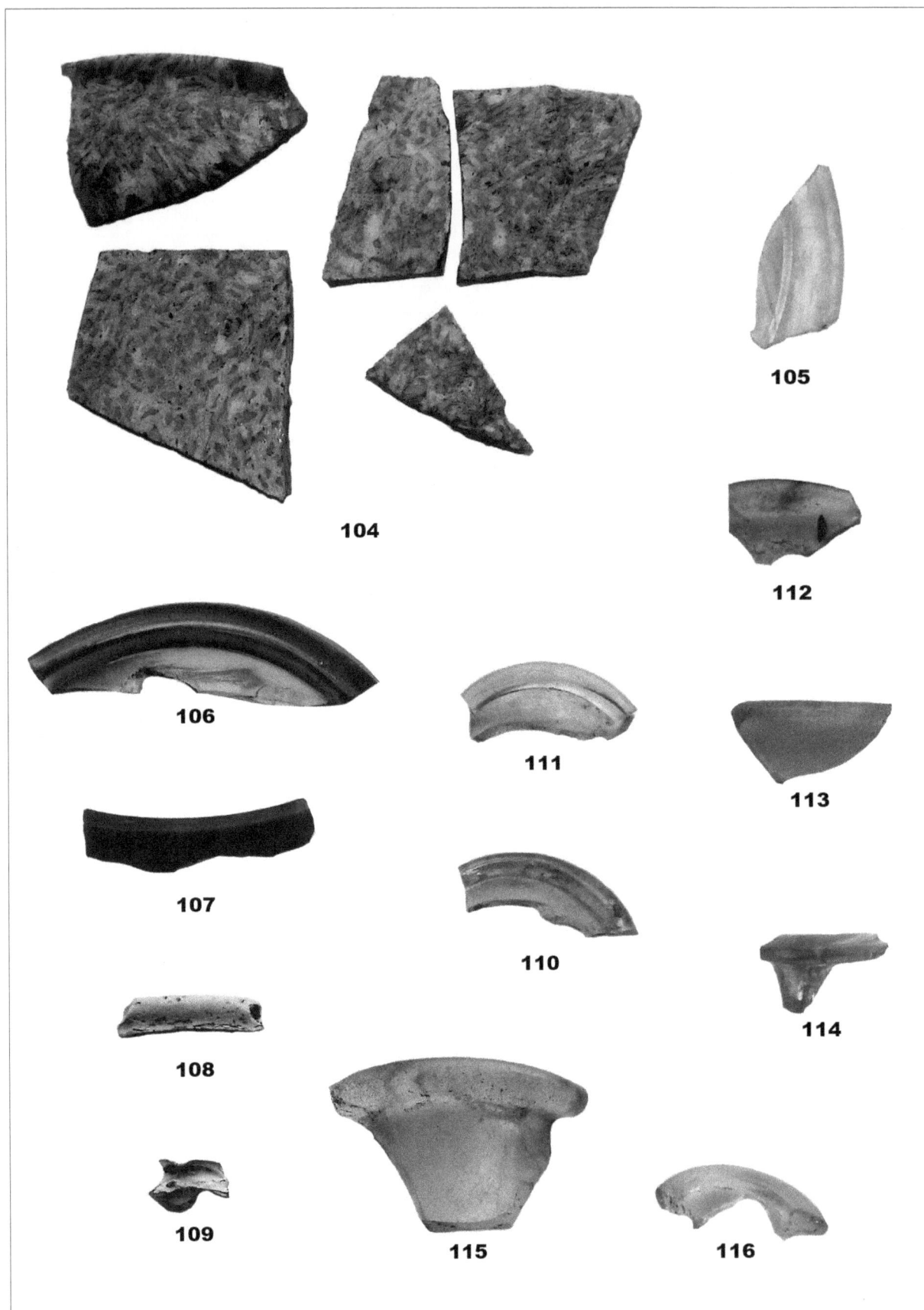

104

105

112

106

111

113

107

110

114

108

109

115

116

Glass. Fig. 104: fragments of a millefiori dish. Figs. 105, 106: fragments of dishes. Figs. 107 - 116: rim fragments of vessels.

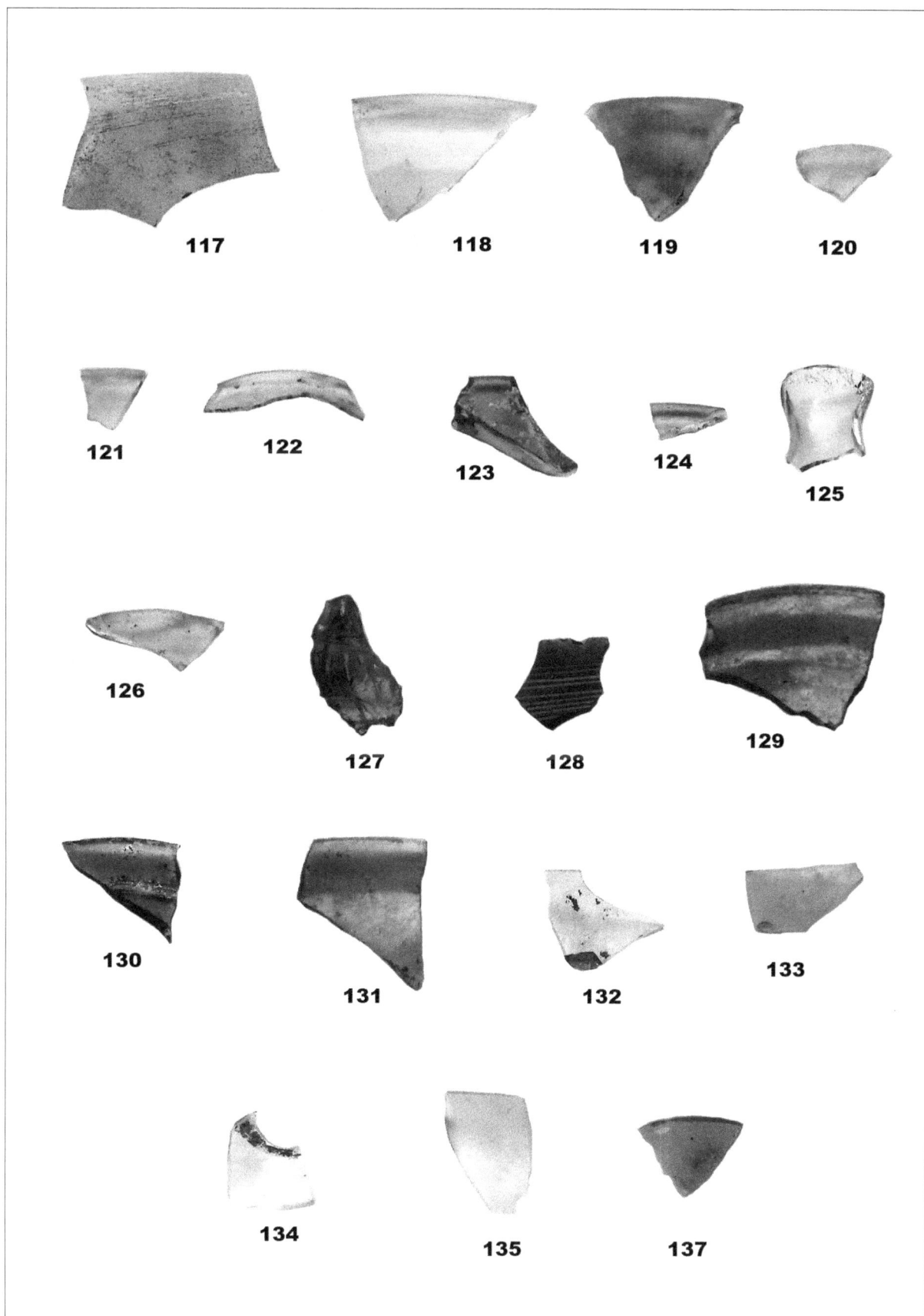

Glass. Figs. 117 - 135, 137: fragments of vessels.

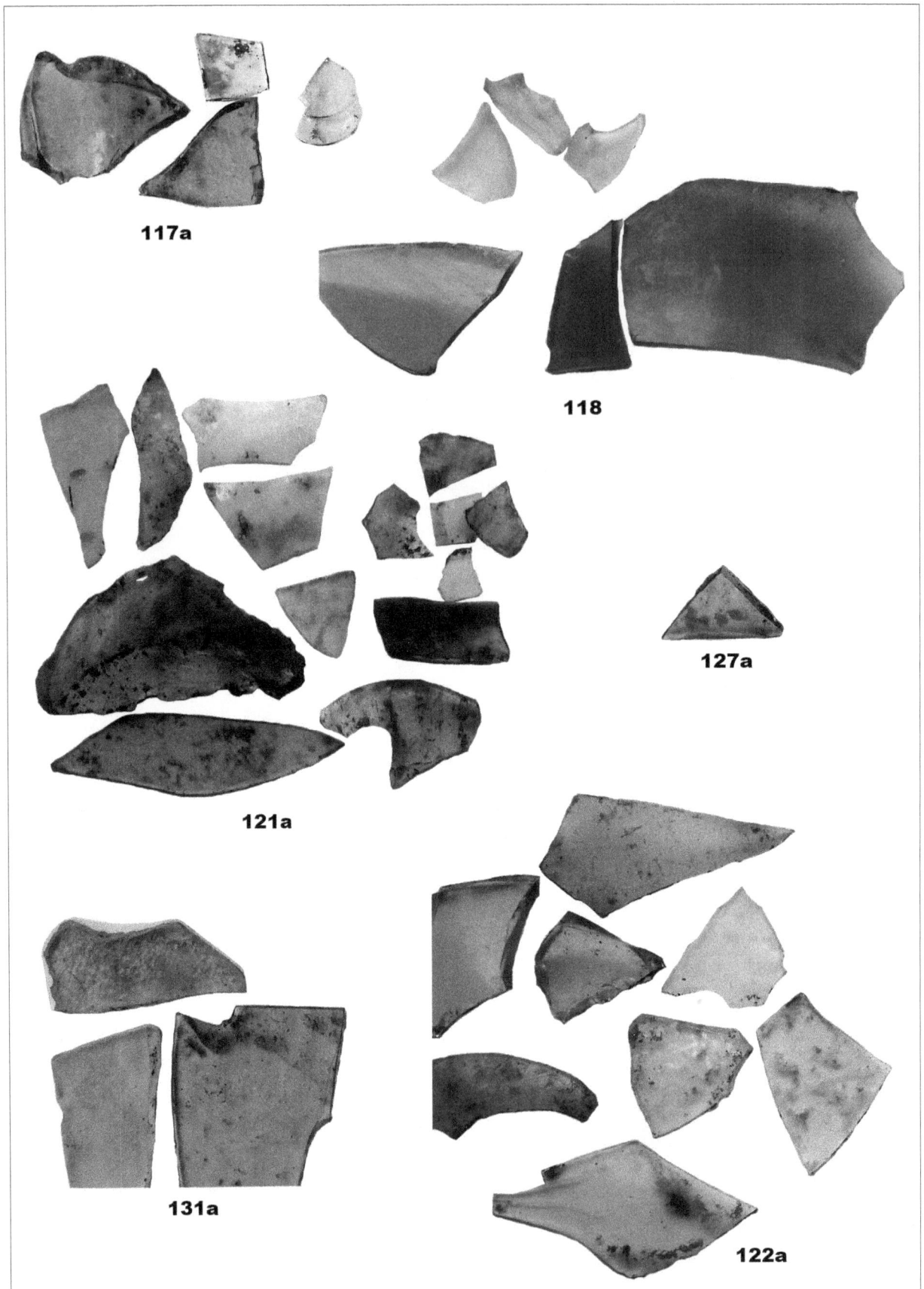

117a

118

121a

127a

131a

122a

Glass. Figs. 117a, 118a, 121a, 122a, 127a, 131a: group of fragments with which items 117, 118, 121, 122, 127, 131 were found.

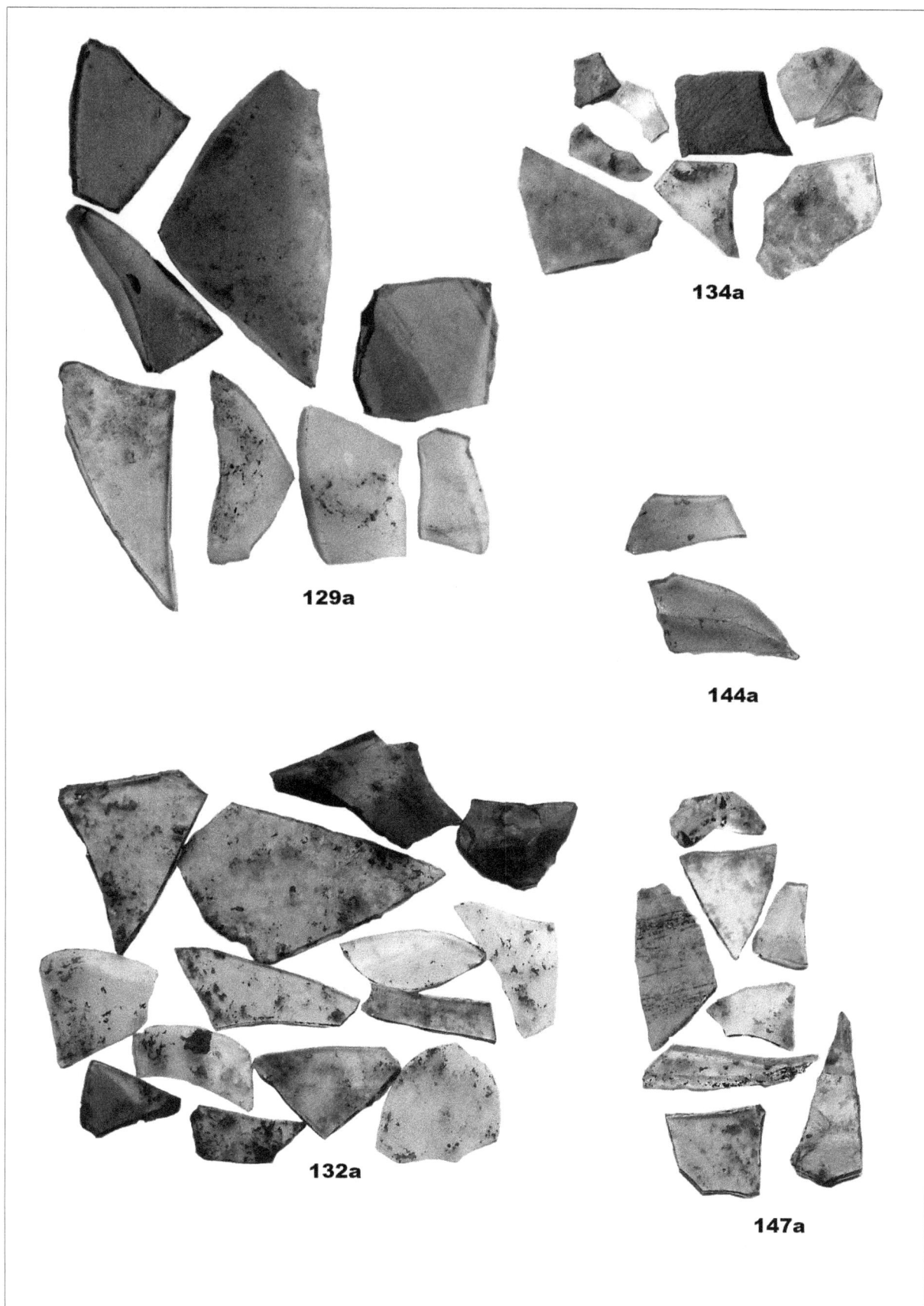

134a

129a

144a

132a

147a

Glass. Figs. 129a, 132a, 134a, 144a, 147a: group of fragments with which items 129, 132, 134, 144, 147 were found.

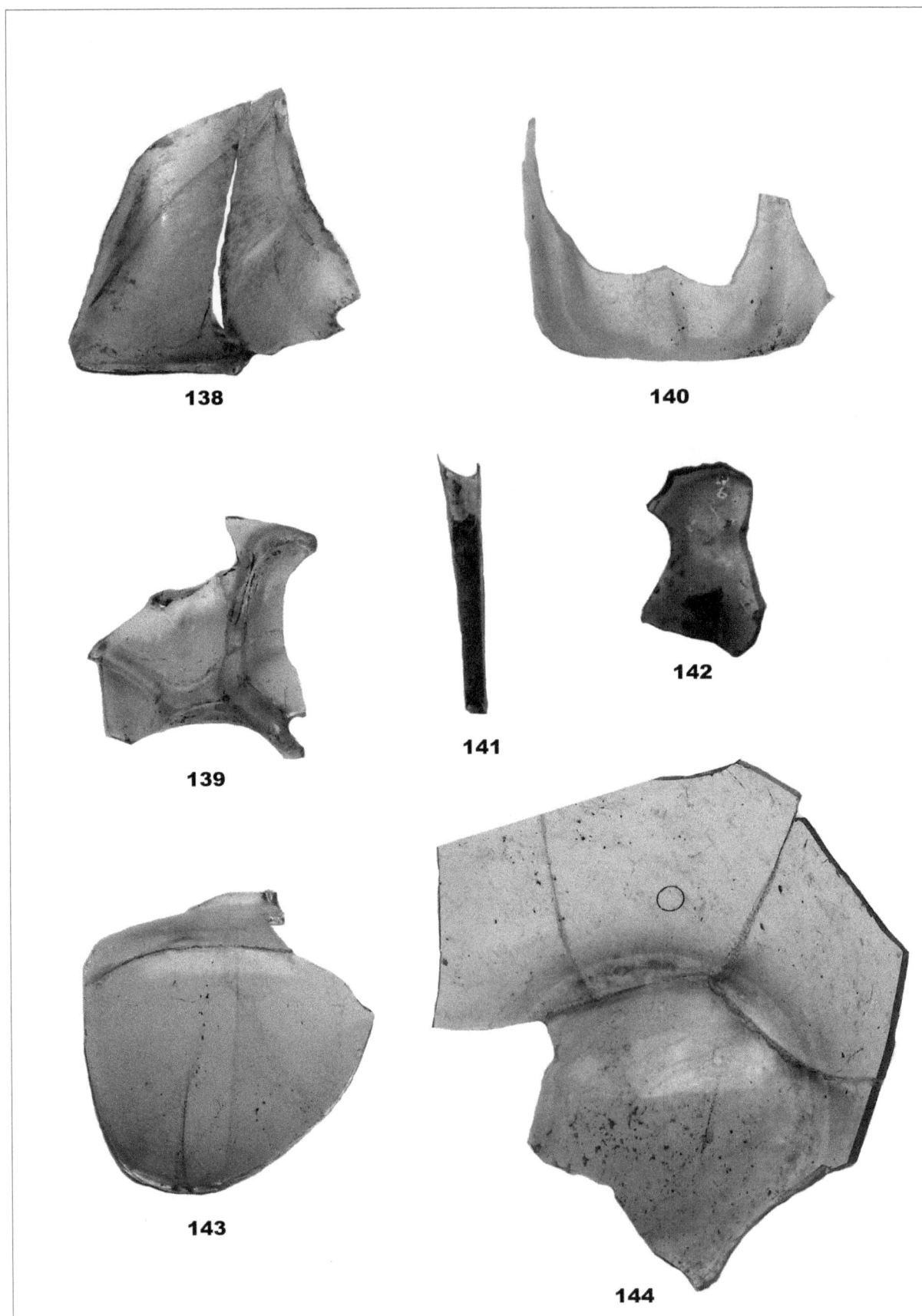

Glass. Figs. 138 - 140, 143, 144: wall and base fragments. Fig. 141: tubular fragment. Fig. 142: wall fragment with small pieces of glass attached to the surface.

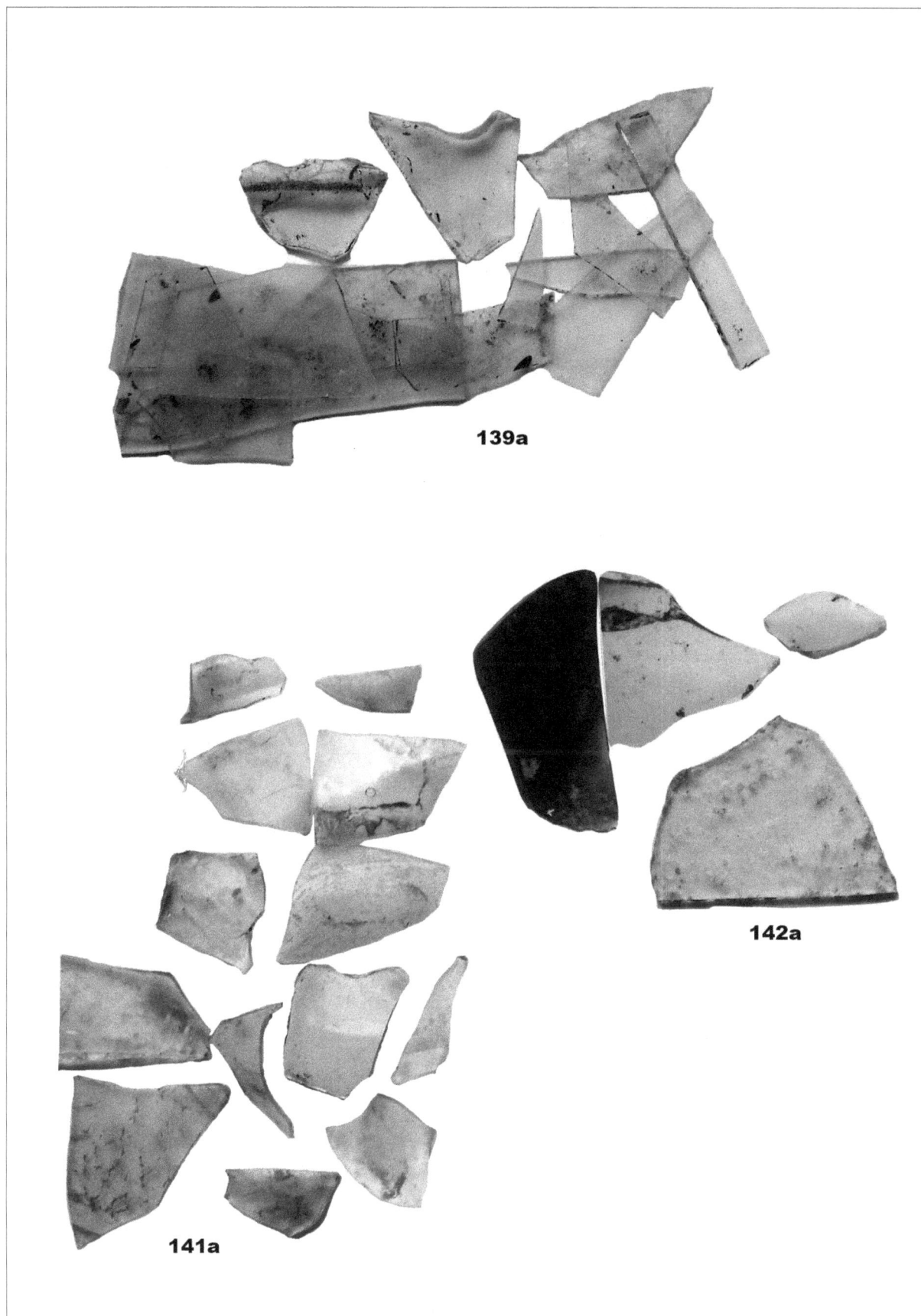

Glass. Figs. 139a, 141a, 142a: group of fragments with which items 139, 141, 142 were found.

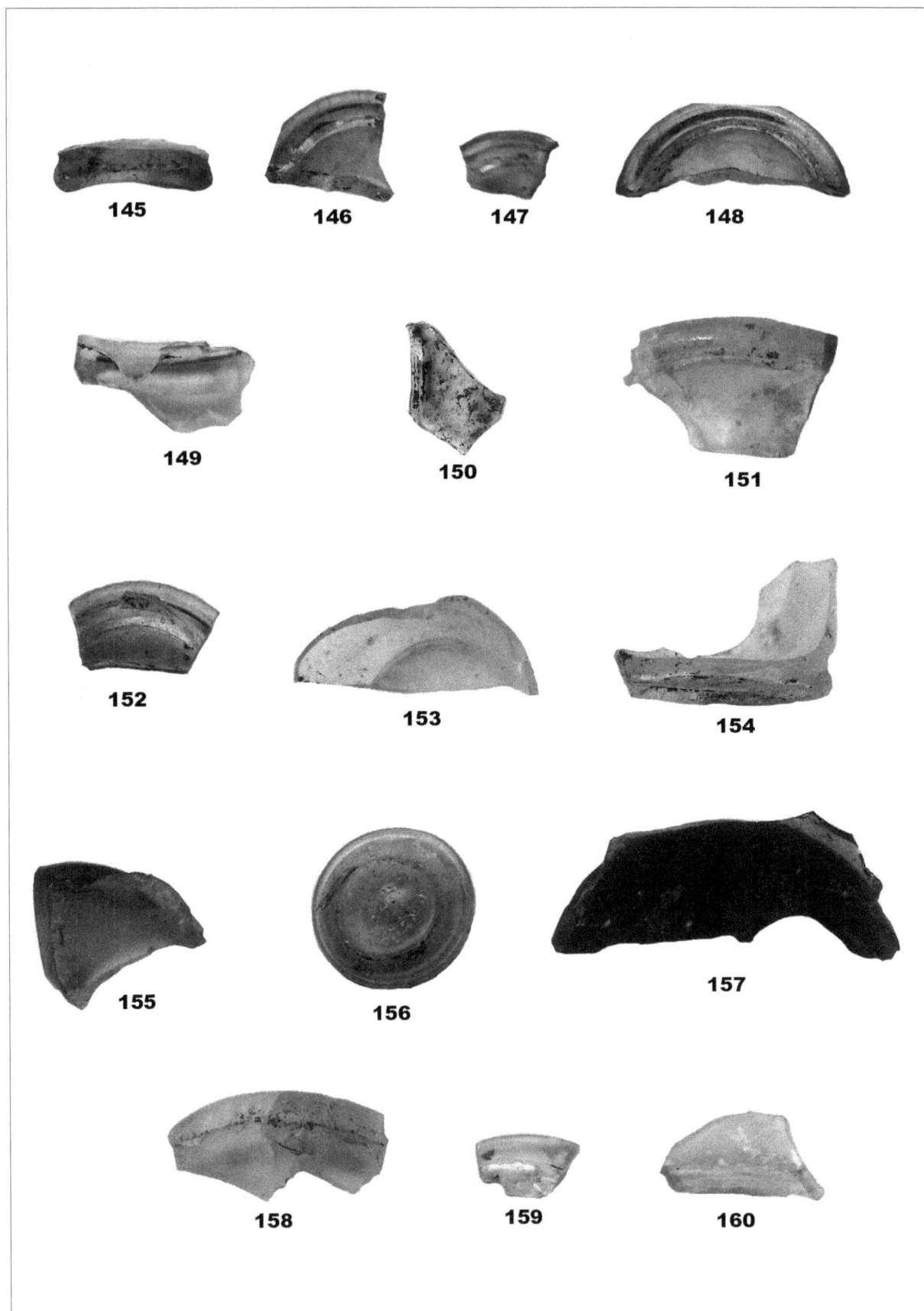

Glass. Figs. 145 - 160: fragments of the lower parts of vessels.

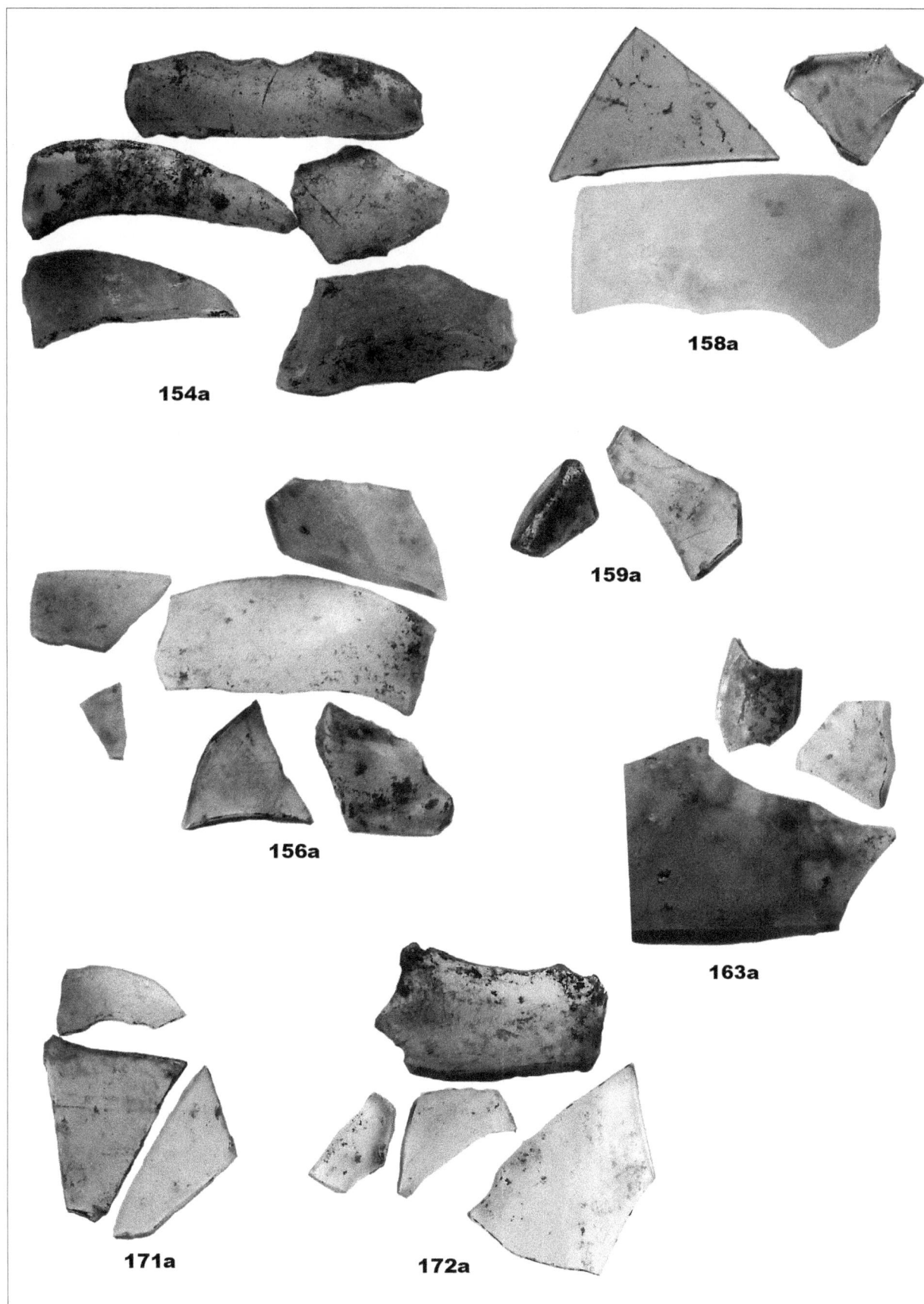

154a

158a

156a

159a

163a

171a

172a

Glass. Figs. 154a, 156a, 158a, 159a, 163a, 171a, 172a: group of fragments with which items 154, 156, 158, 159, 163, 171, 172 were found.

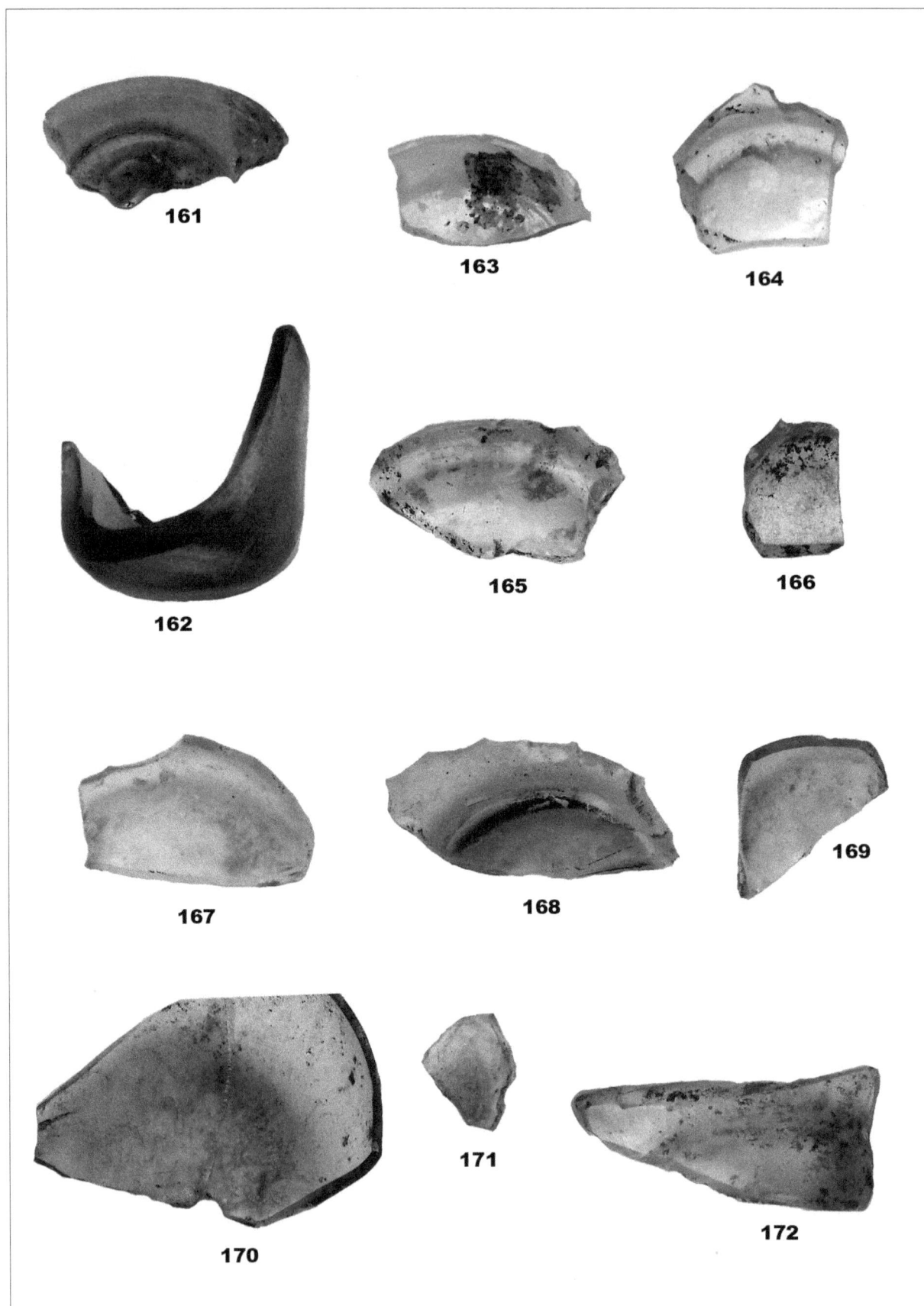

Glass. Figs. 161 - 172: fragments of the lower parts of vessels.

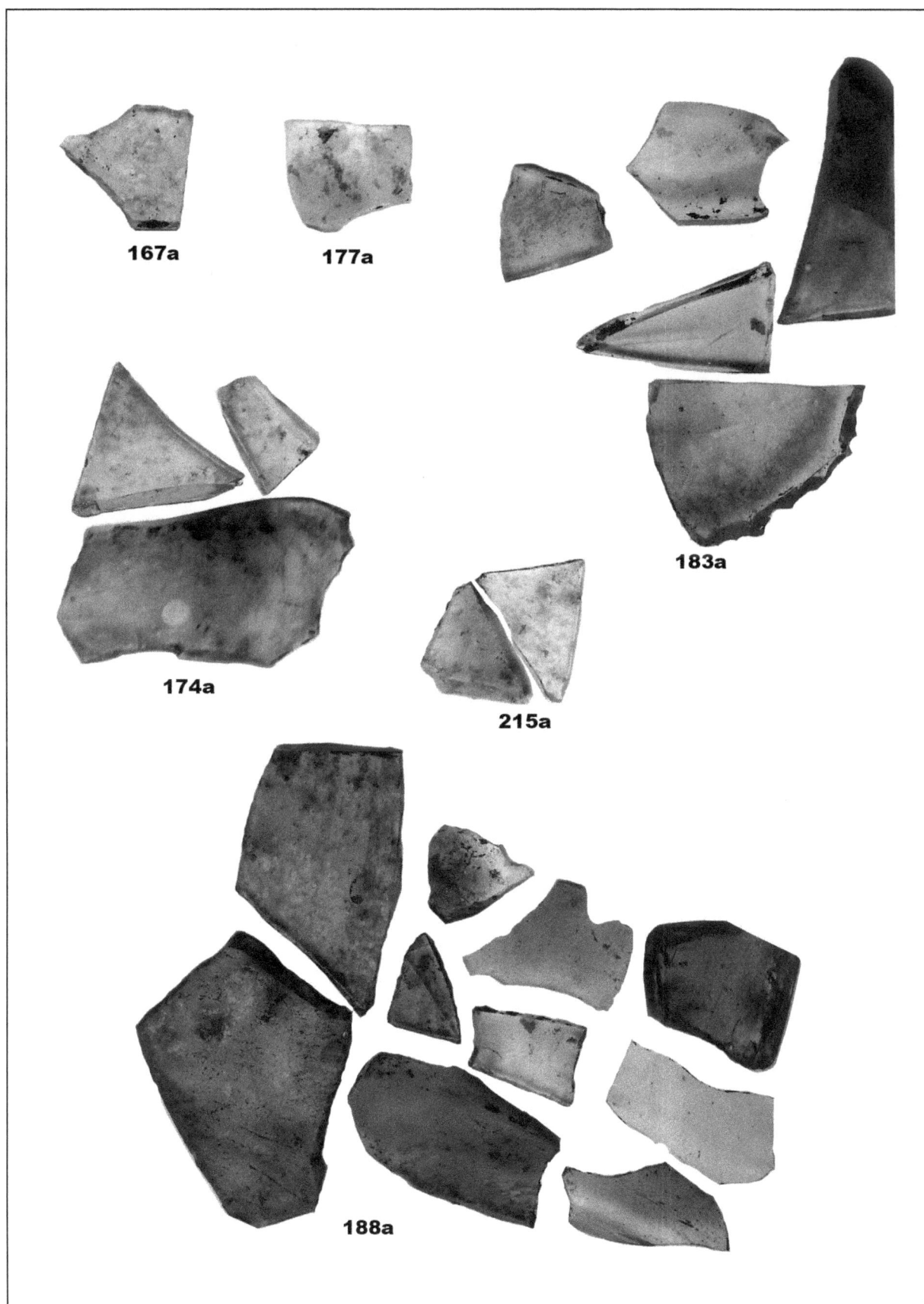

Glass. Figs. 167a, 174a, 177a, 183a, 188a, 215a: individual fragments and groups with which items 167, 174, 177, 183, 188, 215 were found.

173

174

176

175

177

179

178

180

181

182

183

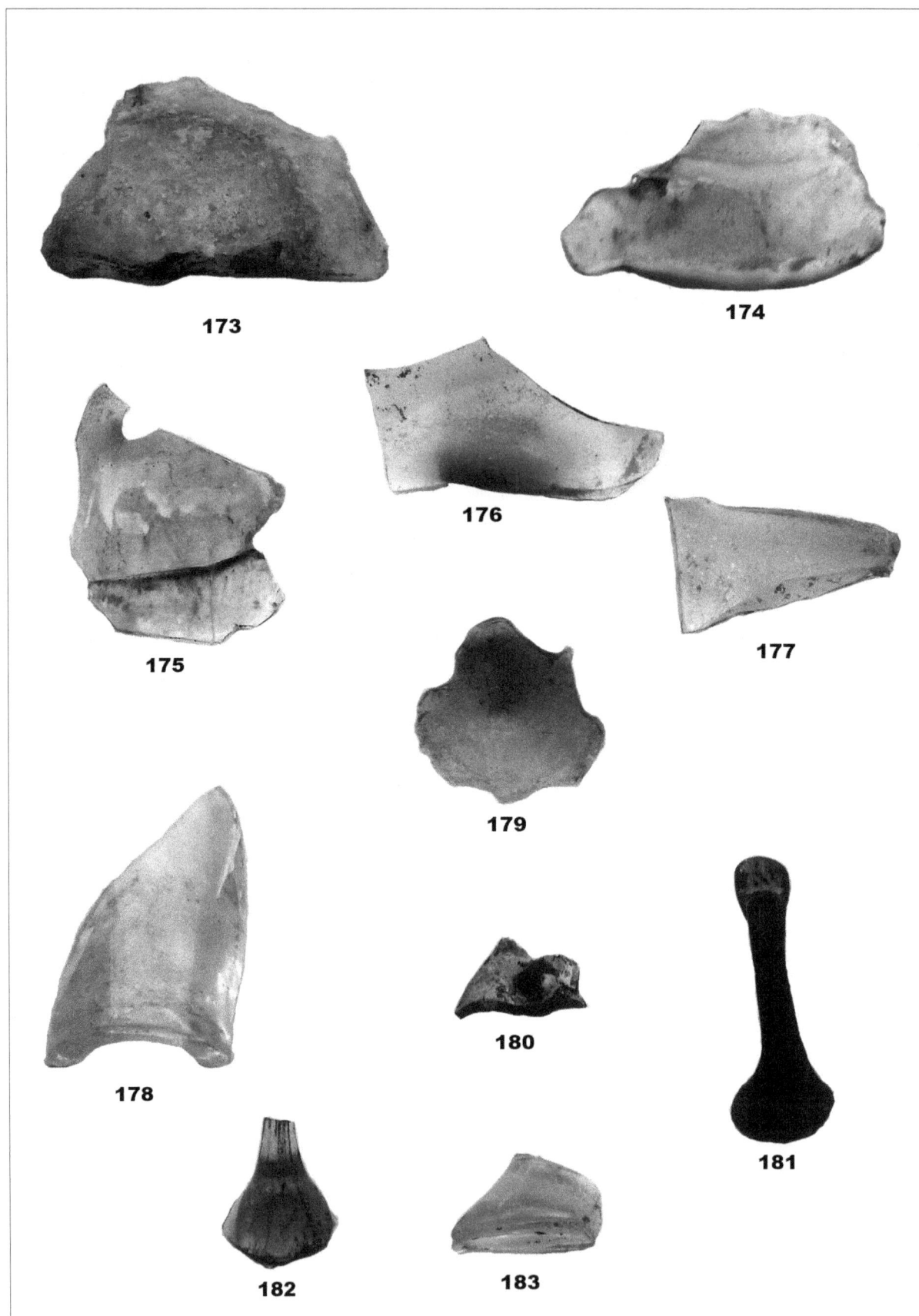

Glass. Figs. 173 - 180: fragments of the lower parts of vessels. Fig. 181: jug handle. Figs. 182, 183: lower sections of handles.

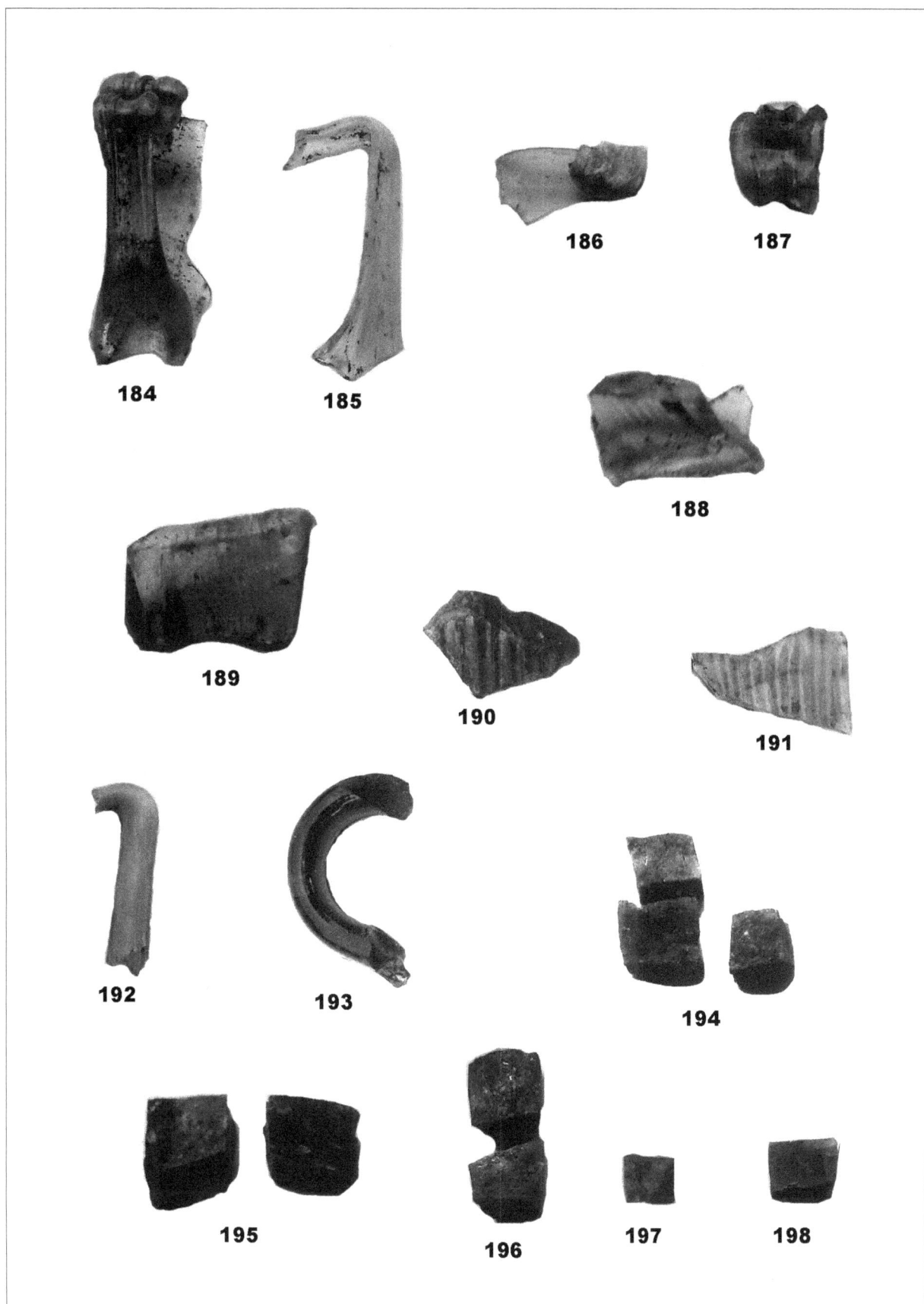

Glass. Figs. 184 - 191: handles and fragments of jugs and flasks. Figs. 192, 193: handles (?) of vessels. Figs. 194 - 198: cubes.

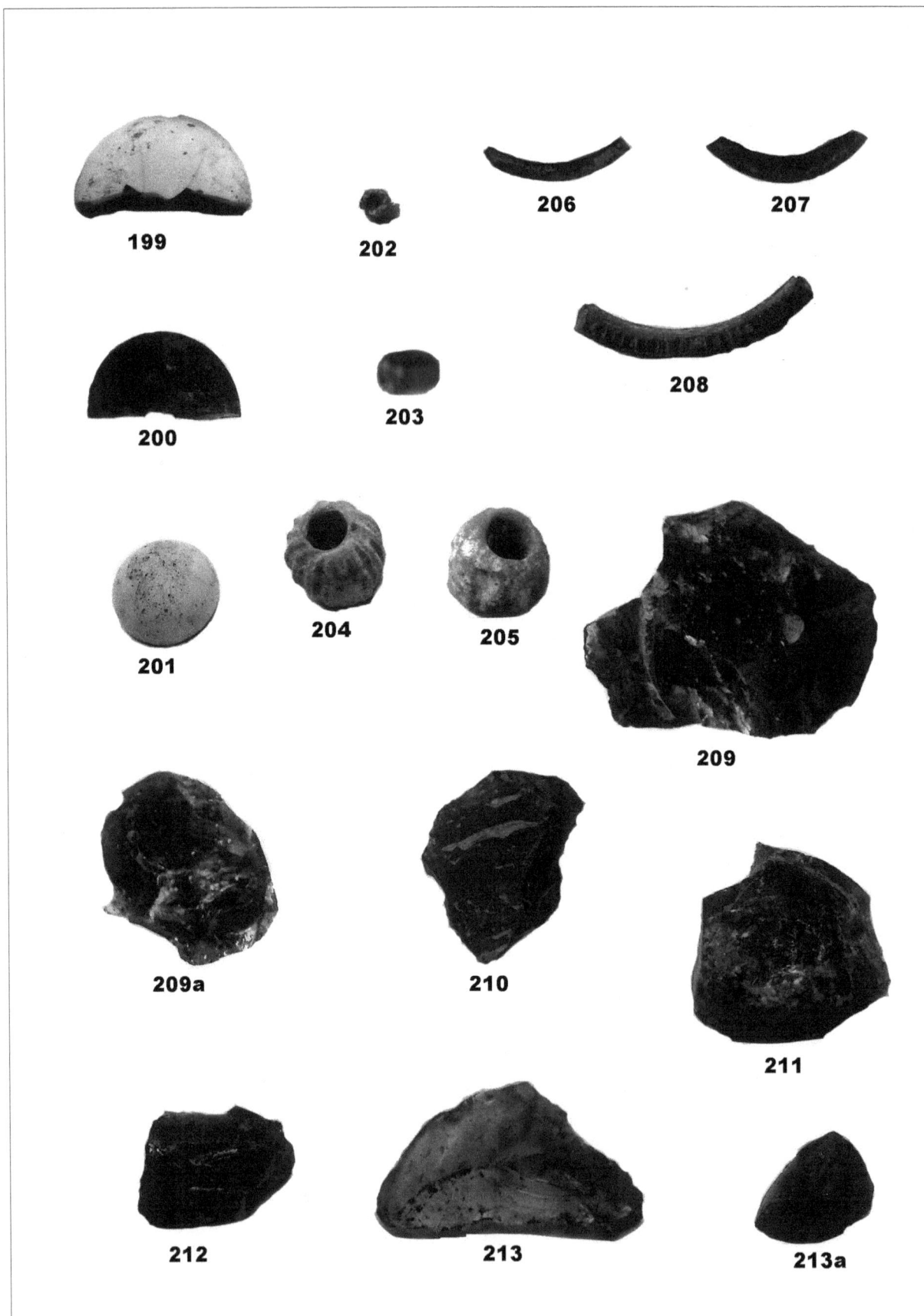

Glass. Figs. 199 - 201: flat circular pieces. Figs. 202 – 205: beads. Figs. 206, 207: bracelet fragments. Figs. 209 - 213a: raw glass. Jet. Fig. 208: part of a bracelet.

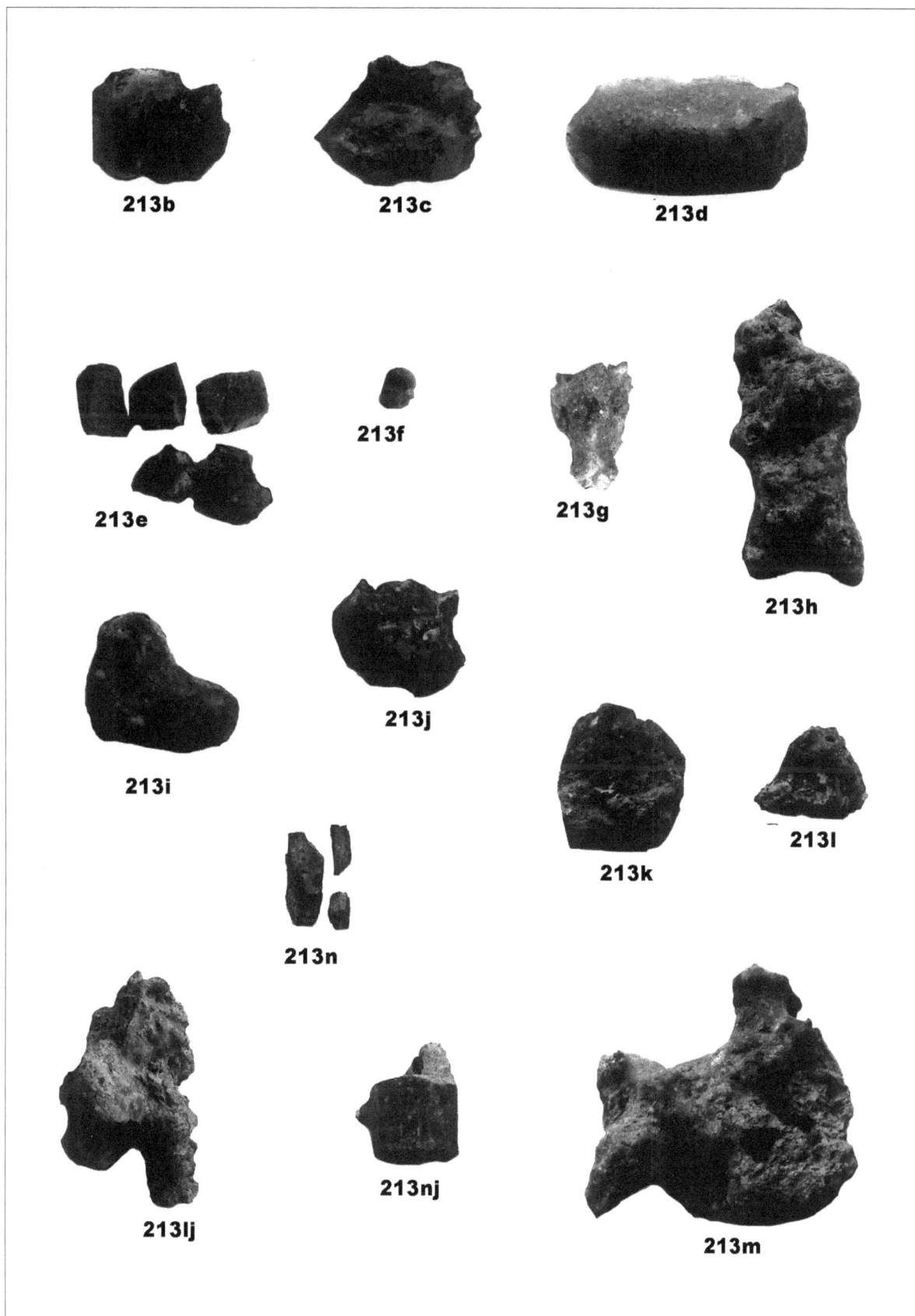

Glass. Figs. 213b - 213e: raw glass. Fig. 213f: drop-shaped clump. Figs. 213g - 213i, 213k, 213l: glass slag. Figs. 213j, 213lj, 213m: glass slag. Figs. 213n, 213nj: pieces of carbonized wood and charcoal.

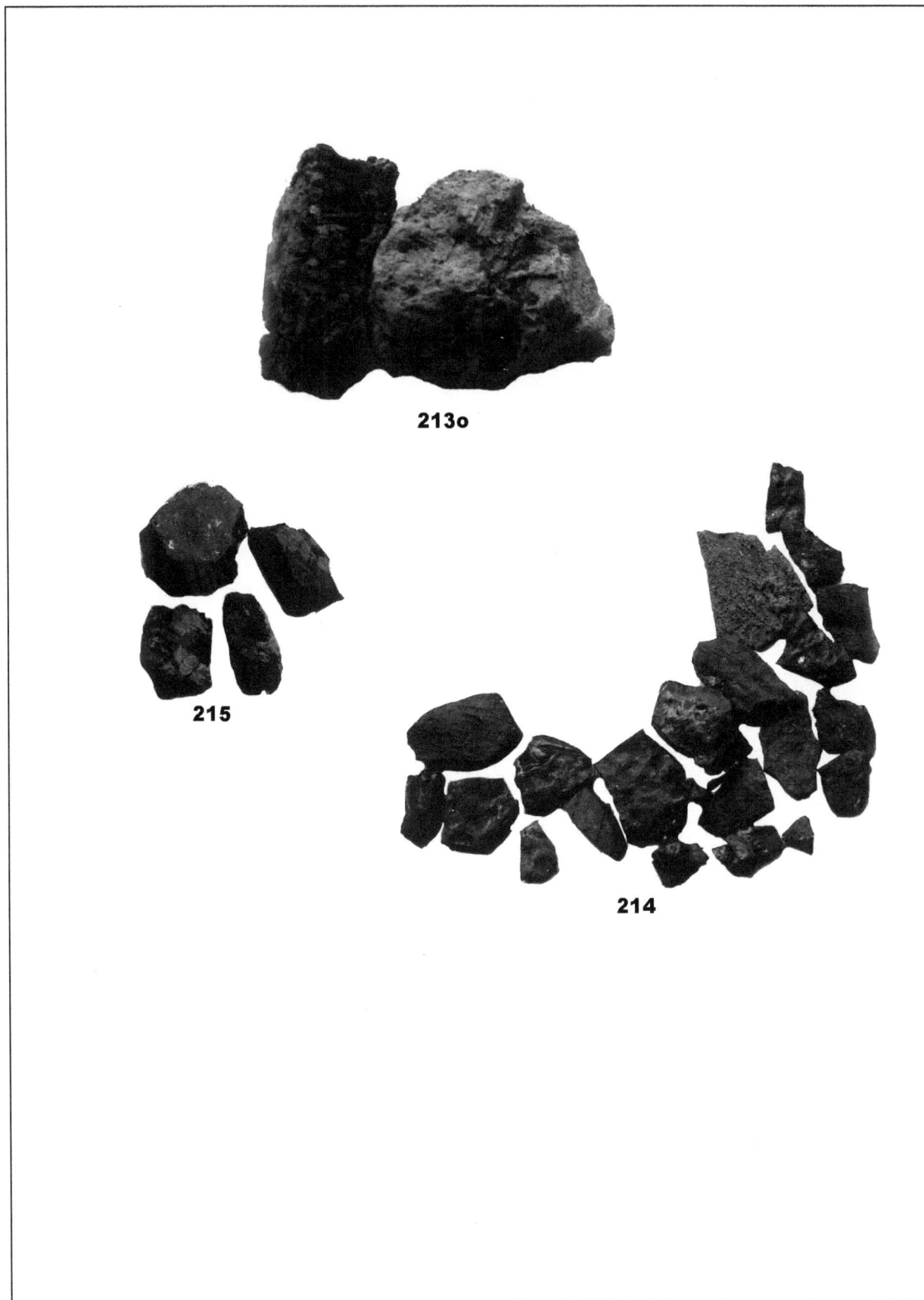

Fig. 213o: a piece of unburnt wood and charcoal. Figs. 214, 215: small pieces of jet.

Glass. Fig. 216: fragments of a vessel.

www.ingramcontent.com/pod-product-compliance
Lightning Source LLC
Chambersburg PA
CBHW061010030426
42334CB00033B/3432